SIGNAL PROCESSING DRIVEN MACHINE LEARNING TECHNIQUES FOR CARDIOVASCULAR DATA PROCESSING

SIGNAL PROCESSING DRIVEN MACHINE LEARNING TECHNIQUES FOR CARDIOVASCULAR DATA PROCESSING

Edited by

RAJESH KUMAR TRIPATHY
Department of Electrical & Electronics Engineering
BITS Pilani Hyderabad
Hyderabad, Telangana, India

RAM BILAS PACHORI
Department of Electrical Engineering, IIT Indore
Indore, Madhya Pradesh, India

ACADEMIC PRESS
An imprint of Elsevier

Academic Press is an imprint of Elsevier
125 London Wall, London EC2Y 5AS, United Kingdom
525 B Street, Suite 1650, San Diego, CA 92101, United States
50 Hampshire Street, 5th Floor, Cambridge, MA 02139, United States

ISBN: 978-0-443-14141-6

For information on all Academic Press publications
visit our website at https://www.elsevier.com/books-and-journals

Publisher: Mara Conner
Acquisitions Editor: Sonnini Yura
Editorial Project Manager: Isabella Silva
Production Project Manager: Anitha Sivaraj
Cover Designer: Greg Harris

Typeset by VTeX

Working together
to grow libraries in
developing countries

www.elsevier.com • www.bookaid.org

Contents

List of contributors

Shrey Agarwal
Department of Computer Science, University of Southern California, Los Angeles, CA, United States

Monika Agrawal
Centre of Applied Research in Electronics, IIT Delhi, New Delhi, India

Chhaviraj Chauhan
Bharti School of Telecommunication Technology and Management, IIT Delhi, New Delhi, India

Cheuk To Skylar Chung
Cardiovascular Analytics Group, PowerHealth Research Institute, Hong Kong, China

Hari Krishna Damodaran
Department of EEE, Birla Institute of Technology and Science, Pilani, Hyderabad, India

Shaswati Dash
Department of EEE, Birla Institute of Technology and Science, Pilani, Hyderabad, India

José Antonio de la O Serna
Autonomous University of Nuevo Leon, Monterrey, NL, Mexico

Shresth Gupta
Department of ECE, IIIT Naya Raipur, Naya Raipur, Chhattisgarh, India

Tapan Kumar Jain
Department of ECE, Indian Institute of Information Technology (IIIT), Nagpur, India

Haipeng Liu
Centre for Intelligent Healthcare, Coventry University, Coventry, United Kingdom

Hemant Kumar Meena
Malaviya National Institute of Technology (MNIT), Jaipur, Rajasthan, India

Umesh Kumar Naik Mudavath

Department of Electronics and Electrical Engineering, Indian Institute of
Technology Guwahati, North Guwahati, Assam, India

Dhanhanjay Pachori

Department of ECE, Indian Institute of Information Technology (IIIT), Nagpur,
India

Ram Bilas Pachori

Department of Electrical Engineering, Indian Institute of Technology Indore,
Indore, India

Ganapati Panda

Department of ECE, CV Raman Global University, Bhubaneswar, India

Mario R. Arrieta Paternina

National Autonomous University of Mexico (UNAM), Mexico City, Mex., Mexico

Abhay Patwari

Department of EEE, Birla Institute of Technology and Science, Pilani, Hyderabad,
India

Vellaisamy Roy

School of Science and Technology, Hong Kong Metropolitan University,
Hong Kong, China

Abhishek Sharma

Department of ECE, IIIT Naya Raipur, Naya Raipur, Chhattisgarh, India

Ramnivas Sharma

Malaviya National Institute of Technology (MNIT), Jaipur, Rajasthan, India

Anurag Singh

Department of ECE, IIIT Naya Raipur, Naya Raipur, Chhattisgarh, India

Rajesh Kumar Tripathy

Department of EEE, Birla Institute of Technology and Science, Pilani, Hyderabad,
India

Gary Tse

School of Nursing and Health Studies, Hong Kong Metropolitan University,
Hong Kong, China

Yashaswi Upmon
Department of Computer Science Engineering, Kalinga Institute of Industrial Technology, Bhubaneshwar, Orissa, India

Alejandro Zamora-Mendez
Michoacan University of Saint Nicholas of Hidalgo, Morelia, Michoacan, Mexico

Muhammad Zubair
Department of Biomedical Engineering, The Pennsylvania State University, State College, PA, United States

CHAPTER 1

Introduction to cardiovascular signals and automated systems

Dhanhanjay Pachori[a], **Shaswati Dash**[b], **Rajesh Kumar Tripathy**[b], **and Tapan Kumar Jain**[a]

[a]Department of ECE, Indian Institute of Information Technology (IIIT), Nagpur, India
[b]Department of EEE, Birla Institute of Technology and Science, Pilani, Hyderabad, India

Contents

1.1 Heart conduction system and ECG signal

The heart is mainly an autonomous organ, and its function is to provide oxygen-rich blood to the entire body [1]. It comprises four chambers, namely, the left atrium (LA), left ventricle (LV), right atrium (RA), and right ventricle (RV). Similarly, it also contains four valves, namely, the tricuspid valve (TV), mitral valve (MV), pulmonary valve (PV), and aortic valve (AV), which help to prevent the backward flow of blood [2]. The flow of blood happens in the heart in a unidirectional manner. Initially, the heart receives deoxygenated blood through the RA, which passes from the RA to the RV through the TV. The function of the RV is to pump the deoxygenated blood to the lungs for purification through the PV. Similarly,

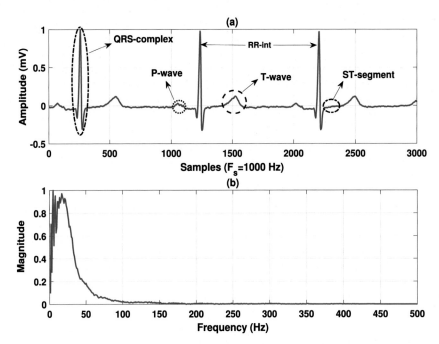

Figure 1.1 (a) ECG signal with marked clinical patterns (the sampling frequency of the ECG signal is 1000 Hz). (b) The Fourier spectrum of the ECG signal.

the LA collects the oxygen-rich blood from the lungs, which is passed to the LV. The function of the LV is to pump the oxygenated blood through the AV to the aorta, which further transfers the blood to the entire body [2]. The heart thus acts as a two-way pump for transferring blood. The heart's function is based on an electrical conduction path. The pacemaker cells, also known as cardiac myocytes, establish the conduction path. These pacemaker cells form the (a) sino–atrial (SA) node, (b) atrioventricular (AV) node, (c) His bundle, and (d) Purkinje fibers [3]. The firing rate of each pacemaker cell is different. The SA and AV nodes have firing rates of 60 to 100 beats per minute (bpm) and 40 to 60 bpm, respectively. Similarly, for the His bundle and Purkinje fibers, the firing rates vary between 20 and 45 bpm [4].

The electrocardiogram (ECG) is a graphical representation of the heart's electrical activity and captures information regarding the depolarization and repolarization of heart chambers. A typical lead I ECG signal for a normal sinus rhythm (NSR) is shown in Fig. 1.1(a). The ECG signal is collected from the PTB diagnostic 12-lead ECG database [5] [6]. It consists of clinical patterns such as the P-wave, QRS-complex, T-wave, and baseline. Features such as duration and amplitude of the clinical patterns (P-wave, QRS-complex, T-wave) are evaluated from the ECG signal [2]. These features are widely used in clinical studies for cardiovascular disease diagnosis and other biomedical applications [7] [8] [9]. Similarly, the Fourier spectrum of the ECG signal is shown in Fig. 1.1(b). It is evident that significant spectral energy is observed between 0.5 Hz and 45 Hz in the spectrum of the ECG signal. The time-domain and spectral characteristics of ECG signals vary for different heart diseases [10]. The details regarding the features of ECG signals are described in the following subsection.

1.1.1 Features of ECG signals

The P-wave represents the electrical activity regarding atrial depolarization [2]. Typically, the amplitude of the P-wave is 0.25 mV for an NSR. Similarly, the duration of the P-wave varies between 80 ms and 100 ms for a normal person. During atrial hypertrophy-based cardiac ailments, the ECG signal has a higher (amplitude more than 0.25 mV) and wider P-wave (duration longer than 100 ms) [2]. Similarly, abnormal and unordered P-waves are seen in the ECG signal during atrial fibrillation (AF) [2]. The QRS-complex reveals the electrical activity for the depolarization of the left and right ventricles of the heart. The QRS-complex duration varies between 80 ms and 100 ms for a normal heart rhythm in healthy individuals. The increase in the duration of QRS-complex beyond 100 ms indicates branch bundle block (BBB) [2]. The ST-segment in the ECG represents the interval between the end of ventricular depolarization and the beginning of ventricular repolarization. Typically, the slope of the ST-segment is used in clinical settings as an important biomarker for the diagnosis of myocardial infarction (MI). The T-wave in the ECG reveals the electrical activity for the repolarization of ventricles. For a healthy heart or NSR, the amplitude and duration of the T-wave are less than 0.5 mV and between 100 ms and 250 ms, respectively [2]. The change in the morphology of the T-wave is an important parameter for the diagnosis of MI and cardiomyopathy-based heart diseases [2].

1.1.2 Heart diseases and morphological changes in ECG signals

Various heart diseases are diagnosed based on morphological changes in the ECG signal [2]. These cardiac diseases include (a) MI, (b) BBB, (c) AF, and (d) ventricular tachycardia (VT) [10]. MI-based cardiac diseases mainly happen due to obstruction in one of the coronary arteries of the heart [11]. This disease progresses into three main stages: the ischemic stage, acute stage, and necrosis stage [8]. The ischemic stage corresponds to the reduction of blood flow in the coronary artery due to atherosclerosis plaque formation [12]. The inverse T-wave elevated ST-segment and abnormal Q-wave morphology are interpreted as the morphological changes in ECG signals due to MI-based heart disease [11]. The ECG signal for the MI case is depicted in Fig. 1.2(a). Similarly, the spectrum of MI pathology-based ECG signal is shown in Fig. 1.2(b). It can be observed that both temporal and spectral information of ECG data changes due to MI pathology as compared to the NSR case in Fig. 1.1(a). Similarly, BBB occurs due to a delay in the heart's conduction system [2]. This type of delay can occur during the depolarization of the heart's ventricles. The morphological changes such as a wider QRS-complex and an increased R-wave amplitude are observed in ECG signals due to BBB-based heart disease. AF is interpreted as an irregular and rapid atrial rhythm, and it can increase the chances of heart failure and MI [13]. The fibrillatory (F)-waves and varying RR intervals are seen in the ECG signal due to AF [14] [2]. The ECG signals for BBB and AF-based heart disease cases are shown in Fig. 1.2(c) and Fig. 1.2(e), respectively. The spectra of the ECG signals for BBB and AF cases are depicted in Fig. 1.2(d) and Fig. 1.2(f), respectively. It is evident that the morphologies of ECG data for AF and BBB differ from the NSR case. There are variations in the spectra of AF- and BBB-based ECG signals [10]. Similarly, VT occurs due to damage in the heart muscle and causes heart failure, or sudden cardiac death [15]. An abnormal QRS-complex with a duration longer than 0.14 s occurs in the ECG signal due to VT [2].

1.1.3 Automated disease diagnosis system using ECG

The continuous monitoring of ECG data from subjects at the intensive care unit (ICU) and wearable devices generates a huge volume of data [16]. It is a time-consuming process for medical staff to manually investigate all morphological features of ECG data in diagnosing cardiac diseases [10]. The manual diagnosis procedure is prone to error, and hence the automated diagnosis system (ADS) is used to detect cardiac diseases from ECG data. The flow chart of the ADS is depicted in Fig. 1.3. It mainly consists of three important stages: (a) recording of ECG data, (b) preprocessing of

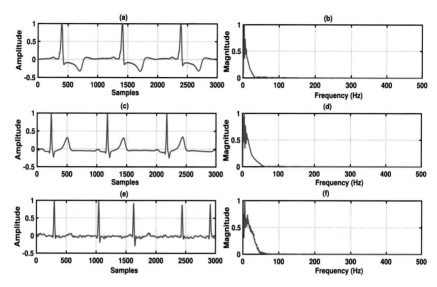

Figure 1.2 (a) ECG signal for MI. (b) The spectrum of MI pathology-based ECG signal. (c) ECG signal for BBB. (d) The spectrum of BBB pathology-based ECG signal. (e) ECG signal for AF. (f) The spectrum of AF pathology-based ECG signal.

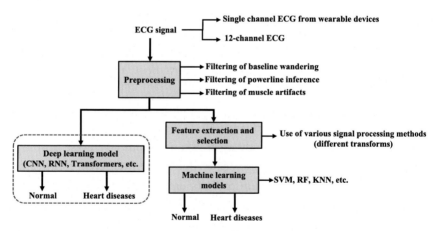

Figure 1.3 Automated diagnosis system for heart disease detection using ECG signals.

ECG data, and (c) the use of signal processing–domain machine learning (ML)- and deep learning (DL)-based methods for automated detection of cardiac diseases [8].

1.1.3.1 Recording of ECG signals

The ECG data can be recorded in the resting state [2] and when the subject is moving [17]. The 12-lead ECG (with lead numbers I, II, III, aVR, aVL, aVF, V1, V2, V3, V4, V5, and V6) is used for the recording of resting-state ECG signals [11]. The 12-lead ECG data provide more information as the heart's electrical activity is viewed from distinct angles. For diagnosing MI, BBB, hypertrophy, and other cardiac arrhythmias, the 12-lead ECG recording is normally recommended by medical professionals [2]. Similarly, the single-channel-based ECG recording is performed in wearable devices for the daily monitoring of heart rate and morphological features of ECG data for different applications like detection of AF [18], stress monitoring [19], and human activity recognition [20]. Furthermore, the ECG signals recorded either in the resting state or from wearable devices are transmitted for telehealthcare monitoring and telemedicine applications [21].

1.1.3.2 Preprocessing of ECG data

In ADS, the preprocessing stage mainly consists of eliminating various types of noises and artifacts from the ECG signals and segmenting ECG data into frames or beats. The ECG data are mainly contaminated with baseline wandering (BW), powerline interference (PI), and muscle or electromyogram noise [22]. The BW noise distorts the below 1 Hz frequency component of ECG data, and it occurs due to body movement, respiration, and improper fixation of electrodes during recording [22]. The ST-segment and isoelectric line parts of ECG data are contaminated due to BW noise. Typically, the discrete wavelet transform (DWT), nonlinear filtering, and other transform-domain-based methods are used to filter out BW noise from the ECG data [22]. Similarly, PI noise occurs due to the interference of nearby equipment during the acquisition of ECG data. Typically the PI noise affects the 50 Hz or 60 Hz frequency and its harmonics in the ECG signal. PI noise can be eliminated using the infinite impulse response (IIR) and adaptive filters [22]. Likewise, muscle noise occurs due to the activity of skeletal muscle, and it is predominant during ambulatory ECG recording or when using wearable ECG recording devices [23]. The muscle noise contaminates the frequency range up to 100 Hz in the ECG data. DWT, eigenanalysis, and DL-based methods are employed to eliminate muscle artifacts from ECG signals [24].

1.1.3.3 ECG feature extraction and selection

ECG feature extraction and feature selection are important stages of designing ADSs to detect cardiac diseases. Two types of methods are normally used to evaluate features from ECG signals. These methods are called direct and indirect methods. In the direct method, the clinical features or morphological features include the amplitude of the P-wave, the maximum positive amplitude of the QRS-complex, the maximum negative amplitude of the QRS-complex, the RR duration, PR duration, and QT duration, the T-wave amplitude, the ST-segment slope, the QRS angle, and the QRS duration. Medical practitioners check the deviation in the aforementioned morphological features from the normal range to diagnose different cardiovascular diseases. To evaluate the morphological features from ECG data, it is required to detect P-, Q-, R-, S-, and T-points. Similarly, in the indirect method, signal processing algorithms such as DWT, eigenspace analysis [11], nonlinear analysis, and spectral analysis-based techniques are used to evaluate the features from ECG signals. The selection of features from all evaluated features from ECG signals is interesting to boost the classification performance of ML algorithms for the automated detection of cardiovascular diseases. Correlation-based feature selection (CFS) [11], mutual information, and other evolutionary computing algorithms have been used in ADSs.

1.1.3.4 Machine learning and deep learning

The detection of cardiac ailments from the feature vectors of ECG signals is performed using ML-based methods. Supervised and unsupervised learning techniques have been used [11] [25]. Under supervised learning, the K-nearest neighbors (KNN), support vector machines (SVM), neural networks, and random forest (RF) have been used to detect cardiac ailments [11] [15]. Similarly, unsupervised learning-based methods such as K-means clustering [25] and hierarchical agglomerative clustering [26] have been utilized to evaluate disease types from the feature vectors of ECG signals.

The DL techniques use the ECG data directly to detect heart diseases [8]. The feature extraction and selection stages are not required in DL-based models for cardiac disease detection using ECG signals. The multiscale-domain [27] and time-frequency-domain DL [13] models are also used to detect different cardiac diseases using single-lead and 12-lead ECG signals. The performance of ML and DL-based models has been evaluated using various performance measures such as accuracy, sensitivity, specificity, kappa score, precision, and F1-score [28].

1.2 Cardiac auscultation and PCG signal

Cardiac auscultation is a simple and noninvasive procedure to listen to the heart sounds or vibrations during the opening and closing of heart valves [28]. The important findings obtained by cardiac auscultation include heart sounds, murmurs, and rubs. The PCG is a digital recording revealing information regarding the vibratory sounds produced during a cardiac cycle [29]. The automated analysis of PCG data is used to diagnose various heart valve diseases (HVDs), heart failure, and atrial and ventricular septal defects. The PCG signal for the normal case and the spectrum of this signal are shown in Fig. 1.4(a) and Fig. 1.4(b), respectively. It consists of the fundamental sounds S1 and S2. The interval between the onset of the S1-sound and the onset of the S2-sound is called the systolic interval [29]. Similarly, the interval between the onset of the S2-sound and the onset of the S1-sound is called the diastolic interval. S1-sound is a low-pitch sound with a longer duration than the S2-sound. The frequency ranges of S1- and S2-sounds are 10–200 Hz and 10–400 Hz, respectively [29]. The other heart sounds, such as S3- and S4-sounds, occur in PCG due to sudden deceleration of blood flow within the ventricle. The S3- and S4-sounds are associated with the diastolic interval region after the S2-sound. These two sounds in PCG are symptoms of heart failure. In pregnant females and children, the presence of the S3-sound is normal [28]. Similarly, murmurs are also seen in PCG signals, and these occur due to the abnormal flow of blood across the heart valves. The murmurs are categorized as systolic, diastolic, and continuous. The frequency ranges for systolic and diastolic murmurs are up to 600 Hz in the PCG signal [30].

1.2.1 Heart valve diseases and changes in PCG

HVDs occur due to damage in heart valves or improper opening and closing of heart valves during the cardiac cycle [28]. The temporal and spectral characteristics of PCG signals vary between HVDs. The plots of PCG signals for aortic stenosis (AS), mitral stenosis (MS), and mitral regurgitation (MR) are depicted in Fig. 1.5(a), Fig. 1.5(c), and Fig. 1.5(e), respectively. Similarly, the spectra of PCG signals for AS, MS, and MR are shown in Fig. 1.5(b), Fig. 1.5(d), and Fig. 1.5(f), respectively. AS occurs due to the narrowing of the aortic valve, which happens due to ventricular hypertrophy [31]. The systolic murmurs are seen in the PCG signal due to AS-based HVD. Similarly, the improper opening or narrowing of the mitral valve causes MS pathology. In the PCG signal, low-pitch murmurs appear in the

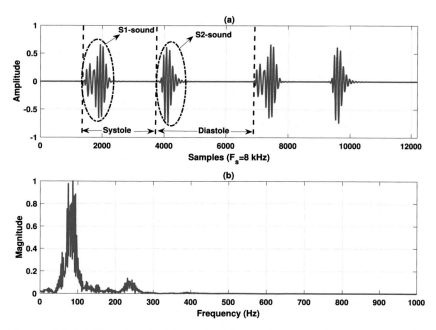

Figure 1.4 (a) PCG signal with marked S1- and S2-sound components (the sampling frequency of the ECG signal is 8000 Hz). (b) The Fourier spectrum of the PCG signal.

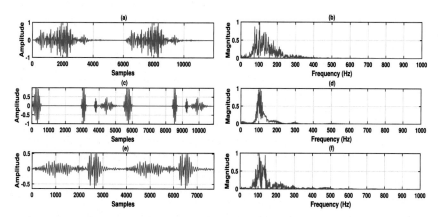

Figure 1.5 (a) PCG signal for aortic stenosis. (b) Spectrum for aortic stenosis pathology. (c) PCG signal for mitral stenosis. (d) Spectrum for mitral stenosis pathology (e) PCG signal for mitral regurgitation. (f) Spectrum for mitral regurgitation pathology.

diastolic region due to MS pathology [32] [28]. Likewise, MR-based HVD occurs due to improper closing of the mitral valve, which further causes the backward flow of blood to the left atrium. The widely split S2-component and soft or buried S1-component are seen in the PCG signal due to MR-based HVD [32]. AR-based HVD occurs when the aortic valve of the heart does not close properly [31]. In the PCG signal, the high-pitch and blowing diastolic murmurs and slapping or sharp S2-sound components are observed due to AR pathology [28].

1.2.2 Automated detection of HVDs using PCG

In clinical settings, the stethoscope is used by medical practitioners to listen to heart sounds for the diagnosis of HVDs [30]. This procedure of diagnosis is manual and requires proper training as well as the listening ability of the medical practitioners for the detection of HVDs. Therefore, automated computer-aided diagnosis frameworks are used for the detection of HVDs using PCG signals [28]. The flowchart of the ADS is shown in Fig. 1.6. The ADS to detect HVDs mainly comprises four steps: filtering of the PCG signal, detection of heart sound activity and segmentation of the PCG signal, extraction of features, and the use of an ML model to detect HVDs. The bandpass filters with a frequency range of 10–800 Hz are typically used for the filtering of recorded PCG data [28]. After filtering, the heart sound activities such as S1-sound and S2-sound and systolic and diastolic regions are extracted from the PCG signal. The feature extraction step involves the evaluation of segmented heart sound audio signal features such as mel frequency cepstral coefficients (MFCCs) and linear prediction coefficients (LPCs). Similarly, various time-frequency-domain analysis methods such as short-time Fourier transform (STFT), continuous wavelet transform (CWT), chirplet transform, and Stockwell transform [28] have been used for the extraction of features from heart sound signals. Once the features from PCG signals or time-frequency representations of PCG data are evaluated, ML- and DL-based techniques are used to detect HVDs [28]. The ML models such as RF, SVM, extreme learning machine, sparse representation classifier, etc. [29] [33], have been used to detect HVDs from PCG signal features. Similarly, the convolutional neural network (CNN) and other DL models have been utilized in the time-frequency representation domain of PCG signals for the automated detection of HVDs [28].

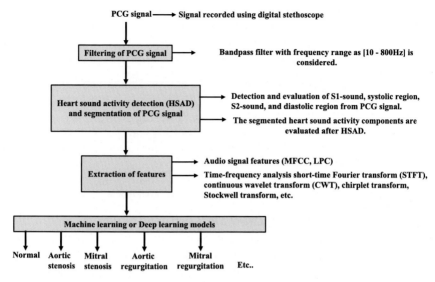

Figure 1.6 Automated diagnosis system for heart valve disease detection using PCG signals.

1.3 PPG signal and cardiorespiratory activity

The PPG signal quantifies the changes in the blood volume during each cardiac cycle, and it is recorded by data acquisition systems (DASs), which consist of a light-emitting diode (LED) transmitter and a photosensitive diode as the reflector [34]. The LED transmitter generates infrared light, illuminating the skin when it falls on the body parts such as the fingertip, forehead, or earlobe. Moreover, the photosensitive diode measures the light absorbed in the body's tissue over time, representing the variations in blood volume [35]. The physiological parameters such as blood oxygen saturation, heart rate, respiratory rate, blood vessel viscosity, blood pressure (BP), and cardiac output can be measured using PPG recordings [36]. The PPG signal contains both pulsatile and superimposed components [37]. The pulsatile component contains information regarding the variations in the blood volume during heartbeats and depends on both the systolic and diastolic phases of the cardiac cycle [36]. Similarly, the superimposed component reveals information regarding respiration, thermoregulation, and activity of the sympathetic nervous system [36]. The frequency ranges for cardiac and respiratory activities are 0.5–3 Hz and 0.14–1 Hz, respectively [38]. The PPG signal and the spectrum of this signal are shown in Fig. 1.7(a) and Fig. 1.7(b), respectively. The PPG signal has been taken from the Real-

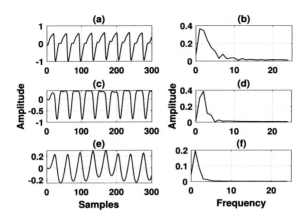

Figure 1.7 Waveforms and spectra of a PPG signal and cardiorespiratory activities of the PPG signal. (a) PPG signal (sampling frequency is $F_s = 50$ Hz). (b) The spectrum of the PPG signal. (c) Cardiac activity component of the PPG signal. (d) The spectrum of the cardiac activity band of the PPG signal. (e) Respiratory activity component of the PPG signal. (f) The spectrum of the respiratory activity band of the PPG signal.

World PPG dataset [39]. The cardiac and respiratory components of PPG signals are depicted in Fig. 1.7(c) and Fig. 1.7(e), respectively; similarly, their spectra are shown in Fig. 1.7(d) and Fig. 1.7(f), respectively.

1.3.1 Automated analysis of PPG signals

The PPG signal is widely used for various cardiovascular data processing applications. These applications include cuffless BP estimation [40], AF detection [41], and heart rate estimation. Gupta et al. [42] proposed a dynamic large artery stiffness index to analyze cuffless BP estimation using the PPG signal. There, the authors evaluated the nonlinear features of the PPG signals using higher-order derivatives and used extreme gradient boosting (XGBoost), random forest (RF), and support vector regressor (SVR) models for the estimation of systolic and diastolic BP values. Filippo et al. [43] used maximal overlap discrete wavelet transform (MODWT)-based features for the estimation of BP. Similarly, Lee et al. [44] used a hybrid feature selection method and estimated the cuffless BP using a decision based on the Gaussian process. In [40], the authors compared 11 ML-based models using PPG signal features to estimate the BP. Cheng et al. [45] proposed the combination of time-frequency analysis along with a DL model to detect the AF using the PPG signal. In [41], the authors reviewed various signal processing, DL, and ML techniques and addressed their disadvantages and future scopes for the detection of AF using PPG data. Furthermore, the authors of [46] used a transfer learning-based approach to precisely detect the AF using multiclass PPG data. In [47], the authors introduced a novel denoising method to eliminate the artifacts from the PPG signal and evaluated the performance of the denoising algorithm based on the estimation of the heart rate from the filtered PPG signals. Similarly, Yang et al. [48] compared the performance of three DL-based models for the accurate estimation of the heart rate from the PPG signals. In [49], the authors used a bandpass filter to extract the time-frequency and statistical features from the PPG signal and used a convolutional recurrent regressor network to estimate the heart rate.

1.4 Future scope of cardiac data processing

The ECG, PPG, and PCG signals are nonstationary; therefore, various nonstationary signal processing techniques can be used to analyze and decompose these signals [50]. Fourier–Bessel series expansion (FBSE) is a nonstationary signal analysis technique which can be used for the analysis of ECG, PPG, and PCG signals [51]. New time-frequency analysis methods such as FBSE-Stockwell transform [52] and eigenvalue decomposition of Hankel matrix with Hilbert transform (EVDHM-HT) [53] can also be used for the analysis and evaluation of time-frequency representations of various cardiac signals. The DL and transfer learning models can be used in the time-frequency representation domain of cardiac signals for the automated detection of heart diseases [14] [8]. The existing methods consider the cardiac signals from one modality for the automated diagnosis of heart diseases and other biomedical signal processing steps. Multisensor-based cardiac signals could be considered for cardiac health monitoring. Multivariate signal processing-based methods such as multivariate empirical wavelet transform (EWT) [54], multivariate projection-based EWT [55], and multivariate fast and adaptive empirical mode decomposition (MFAEMD) [56] can be used for the analysis and decomposition of multichannel- and multisensor-based cardiac signals. The real-time implementation or the hardware design of different signal processing and ML techniques for cardiac health monitoring is a challenging research direction in biomedical engineering and the internet of medical things [57].

1.5 Conclusion

The cardiac signals such as ECG, PCG, and PPG and the information associated with these physiological signals have been discussed in this chapter. The frequency-domain characteristics of ECG, PCG, and PPG data for normal and various abnormal cases (heart diseases) have also been shown. The stages of automated diagnostic systems based on ECG, PCG, and PPG signals have been discussed. The future scope of signal processing and ML techniques in cardiac data processing has been discussed. The development and evaluation of novel features and biomarker-based variations of temporal and spectral information of PCG, ECG, and PPG signals will be interesting for the automated detection of cardiovascular diseases.

References

[1] M.R. Boyett, 'And the beat goes on.' The cardiac conduction system: the wiring system of the heart, Experimental Physiology 94 (10) (2009) 1035–1049.

[2] A.L. Goldberger, Z.D. Goldberger, A. Shvilkin, Clinical Electrocardiography: A Simplified Approach E-Book, Elsevier Health Sciences, 2017.

[3] S. Burkhard, V. Van Eif, L. Garric, V.M. Christoffels, J. Bakkers, On the evolution of the cardiac pacemaker, Journal of Cardiovascular Development and Disease 4 (2) (2017) 4.

[4] J.E. Hall, Guyton & Hall Physiology Review E-Book, Elsevier Health Sciences, 2020.

[5] R. Bousseljot, D. Kreiseler, A. Schnabel, Nutzung der EKG-Signaldatenbank CARDIODAT der PTB über das Internet, Biomedical Engineering (Biomedizinische Technik) 40 (s1) (1995) 317–318, https://doi.org/10.1515/bmte.1995.40.s1.317.

[6] A.L. Goldberger, L.A. Amaral, L. Glass, J.M. Hausdorff, P.C. Ivanov, R.G. Mark, J.E. Mietus, G.B. Moody, C.-K. Peng, H.E. Stanley, Physiobank, physiotoolkit, and physionet: components of a new research resource for complex physiologic signals, Circulation 101 (23) (2000) e215–e220.

[7] F.I. Alarsan, M. Younes, Analysis and classification of heart diseases using heartbeat features and machine learning algorithms, Journal of Big Data 6 (1) (2019) 1–15.

[8] S. Bhaskarpandit, A. Gade, S. Dash, D.K. Dash, R.K. Tripathy, R.B. Pachori, Detection of myocardial infarction from 12-lead ECG trace images using eigendomain deep representation learning, IEEE Transactions on Instrumentation and Measurement 72 (2023) 1–12.

[9] R. Tripathy, Application of intrinsic band function technique for automated detection of sleep apnea using HRV and EDR signals, Biocybernetics and Biomedical Engineering 38 (1) (2018) 136–144.

[10] R.K. Tripathy, S. Dandapat, Automated detection of heart ailments from 12-lead ECG using complex wavelet sub-band bi-spectrum features, Healthcare Technology Letters 4 (2) (2017) 57–63.

[11] L. Sharma, R. Tripathy, S. Dandapat, Multiscale energy and eigenspace approach to detection and localization of myocardial infarction, IEEE Transactions on Biomedical Engineering 62 (7) (2015) 1827–1837.

[12] R. Tripathy, S. Dandapat, Detection of myocardial infarction from vectorcardiogram using relevance vector machine, Signal, Image and Video Processing 11 (6) (2017) 1139–1146.

[13] T. Radhakrishnan, J. Karhade, S.K. Ghosh, P.R. Muduli, R. Tripathy, U.R. Acharya, AFCNNet: automated detection of AF using chirplet transform and deep convolutional bidirectional long short term memory network with ECG signals, Computers in Biology and Medicine 137 (2021) 104783.

[14] H. Manda, S. Dash, R.K. Tripathy, Time-frequency domain modified vision transformer model for detection of atrial fibrillation using multi-lead ECG signals, in: 2023 National Conference on Communications (NCC), IEEE, 2023, pp. 1–5.

[15] R. Tripathy, L. Sharma, S. Dandapat, Detection of shockable ventricular arrhythmia using variational mode decomposition, Journal of Medical Systems 40 (2016) 1–13.

[16] H.M.T. Van, N. Van Hao, K.P.N. Quoc, H.B. Hai, L.M. Yen, P.T.H. Nhat, H.T.H. Duong, D.B. Thuy, T. Zhu, H. Greeff, et al., Vital sign monitoring using wearable devices in a Vietnamese intensive care unit, BMJ Innovations 7 (Suppl 1) (2021).

[17] M. Shao, Z. Zhou, G. Bin, Y. Bai, S. Wu, A wearable electrocardiogram telemonitoring system for atrial fibrillation detection, Sensors 20 (3) (2020) 606.

[18] I.A. Marsili, L. Biasiolli, M. Masè, A. Adami, A.O. Andrighetti, F. Ravelli, G. Nollo, Implementation and validation of real-time algorithms for atrial fibrillation detection on a wearable ECG device, Computers in Biology and Medicine 116 (2020) 103540.

[19] K.M. Dalmeida, G.L. Masala, HRV features as viable physiological markers for stress detection using wearable devices, Sensors 21 (8) (2021) 2873.

[20] F. Attal, S. Mohammed, M. Dedabrishvili, F. Chamroukhi, L. Oukhellou, Y. Amirat, Physical human activity recognition using wearable sensors, Sensors 15 (12) (2015) 31314–31338.

[21] T.H. Falk, M. Maier, et al., MS-QI: a modulation spectrum-based ECG quality index for telehealth applications, IEEE Transactions on Biomedical Engineering 63 (8) (2014) 1613–1622.

[22] R.M. Rangayyan, Biomedical Signal Analysis, John Wiley & Sons, 2015.

[23] S. Luo, P. Johnston, A review of electrocardiogram filtering, Journal of Electrocardiology 43 (6) (2010) 486–496.

[24] W. Mumtaz, S. Rasheed, A. Irfan, Review of challenges associated with the EEG artifact removal methods, Biomedical Signal Processing and Control 68 (2021) 102741.

[25] M. Kaur, A. Arora, Unsupervised analysis of arrhythmias using K-means clustering, IJCSIT International Journal of Computer Science and Information Technologies 1 (5) (2010) 417–419.

[26] H. He, Y. Tan, Automatic pattern recognition of ECG signals using entropy-based adaptive dimensionality reduction and clustering, Applied Soft Computing 55 (2017) 238–252.

[27] R. Panda, S. Jain, R. Tripathy, U.R. Acharya, Detection of shockable ventricular cardiac arrhythmias from ECG signals using FFREWT filter-bank and deep convolutional neural network, Computers in Biology and Medicine 124 (2020) 103939.

[28] J. Karhade, S. Dash, S.K. Ghosh, D.K. Dash, R.K. Tripathy, Time–frequency-domain deep learning framework for the automated detection of heart valve disorders using PCG signals, IEEE Transactions on Instrumentation and Measurement 71 (2022) 1–11.

[29] S.K. Ghosh, R. Ponnalagu, R. Tripathy, U.R. Acharya, Deep layer kernel sparse representation network for the detection of heart valve ailments from the time-frequency representation of PCG recordings, BioMed Research International 2020 (2020).

[30] S.K. Ghosh, R. Ponnalagu, R.K. Tripathy, G. Panda, R.B. Pachori, Automated heart sound activity detection from PCG signal using time–frequency-domain deep neural network, IEEE Transactions on Instrumentation and Measurement 71 (2022) 1–10.

[31] S. Coffey, B.J. Cairns, B. Iung, The modern epidemiology of heart valve disease, Heart 102 (1) (2016) 75–85.

[32] I. Turkoglu, A. Arslan, E. Ilkay, An expert system for diagnosis of the heart valve diseases, Expert Systems with Applications 23 (3) (2002) 229–236.

[33] R.K. Tripathy, S. Dash, A. Rath, G. Panda, R.B. Pachori, Automated detection of pulmonary diseases from lung sound signals using fixed-boundary-based empirical wavelet transform, IEEE Sensors Letters 6 (5) (2022) 1–4.

[34] D. Jarchi, A.J. Casson, Description of a database containing wrist PPG signals recorded during physical exercise with both accelerometer and gyroscope measures of motion, Data 2 (1) (2016) 1.

[35] A. Hadi, Y.H.M. Amin, Designing and constructing an optical monitoring system of blood supply to tissues under pressure, Journal of Medical Signals and Sensors 2 (2) (2012) 114.

[36] A. Prabha, J. Yadav, A. Rani, V. Singh, Intelligent estimation of blood glucose level using wristband PPG signal and physiological parameters, Biomedical Signal Processing and Control 78 (2022) 103876.

[37] Y. Maeda, M. Sekine, T. Tamura, Relationship between measurement site and motion artifacts in wearable reflected photoplethysmography, Journal of Medical Systems 35 (2011) 969–976.

[38] D. Martin-Martinez, P. Casaseca-de-la Higuera, M. Martin-Fernandez, C. Alberola-López, Stochastic modeling of the PPG signal: a synthesis-by-analysis approach with applications, IEEE Transactions on Biomedical Engineering 60 (9) (2013) 2432–2441.

[39] A. Siam, F. Abd El-Samie, A. Abu Elazm, N. El-Bahnasawy, G. Elbanby, Real-world PPG dataset, Mendeley Data 10 (2019).

[40] S. González, W.-T. Hsieh, T.P.-C. Chen, A benchmark for machine-learning based non-invasive blood pressure estimation using photoplethysmogram, Scientific Data 10 (1) (2023) 149.

[41] T. Pereira, N. Tran, K. Gadhoumi, M.M. Pelter, D.H. Do, R.J. Lee, R. Colorado, K. Meisel, X. Hu, Photoplethysmography based atrial fibrillation detection: a review, npj Digital Medicine 3 (1) (2020) 3.

[42] S. Gupta, A. Singh, A. Sharma, Dynamic large artery stiffness index for cuffless blood pressure estimation, IEEE Sensors Letters 6 (3) (2022) 1–4.

[43] F. Attivissimo, L. De Palma, A. Di Nisio, M. Scarpetta, A.M.L. Lanzolla, Photoplethysmography signal wavelet enhancement and novel features selection for non-invasive cuff-less blood pressure monitoring, Sensors 23 (4) (2023) 2321.

[44] S. Lee, G.P. Joshi, A.P. Shrestha, C.-H. Son, G. Lee, Cuffless blood pressure estimation with confidence intervals using hybrid feature selection and decision based on Gaussian process, Applied Sciences 13 (2) (2023) 1221.

[45] P. Cheng, Z. Chen, Q. Li, Q. Gong, J. Zhu, Y. Liang, Atrial fibrillation identification with PPG signals using a combination of time-frequency analysis and deep learning, IEEE Access 8 (2020) 172692–172706.

[46] S. Kudo, Z. Chen, X. Zhou, L.T. Izu, Y. Chen-Izu, X. Zhu, T. Tamura, S. Kanaya, M. Huang, A training pipeline of an arrhythmia classifier for atrial fibrillation detection using photoplethysmography signal, Frontiers in Physiology 14 (2023) 2.

[47] B. Lokendra, G. Puneet, AND-rPPG: a novel denoising-rPPG network for improving remote heart rate estimation, Computers in Biology and Medicine 141 (2022) 105146.

[48] Z. Yang, H. Wang, F. Lu, Assessment of deep learning-based heart rate estimation using remote photoplethysmography under different illuminations, IEEE Transactions on Human-Machine Systems 52 (6) (2022) 1236–1246.

[49] S. Ismail, I. Siddiqi, U. Akram, Heart rate estimation in PPG signals using convolutional-recurrent regressor, Computers in Biology and Medicine 145 (2022) 105470.

[50] R.B. Pachori, Time-Frequency Analysis Techniques and Their Applications, CRC Press, 2023.

[51] P.K. Chaudhary, V. Gupta, R.B. Pachori, Fourier-Bessel representation for signal processing: a review, Digital Signal Processing 135 (2023) 103938.

[52] S. Dash, S.K. Ghosh, R.K. Tripathy, G. Panda, R.B. Pachori, Fourier-Bessel domain based discrete Stockwell transform for the analysis of non-stationary signals, in: 2022 IEEE India Council International Subsections Conference (INDISCON), IEEE, 2022, pp. 1–6.

[53] R.R. Sharma, R.B. Pachori, Eigenvalue decomposition of Hankel matrix-based time-frequency representation for complex signals, Circuits, Systems, and Signal Processing 37 (2018) 3313–3329.

[54] A. Bhattacharyya, R.B. Pachori, A multivariate approach for patient-specific EEG seizure detection using empirical wavelet transform, IEEE Transactions on Biomedical Engineering 64 (9) (2017) 2003–2015.

[55] R.K. Tripathy, S.K. Ghosh, P. Gajbhiye, U.R. Acharya, Development of automated sleep stage classification system using multivariate projection-based fixed boundary empirical wavelet transform and entropy features extracted from multichannel EEG signals, Entropy 22 (10) (2020) 1141.

[56] M.R. Thirumalaisamy, P.J. Ansell, Fast and adaptive empirical mode decomposition for multidimensional, multivariate signals, IEEE Signal Processing Letters 25 (10) (2018) 1550–1554.

[57] A.A. Abdellatif, A. Mohamed, C.F. Chiasserini, M. Tlili, A. Erbad, Edge computing for smart health: context-aware approaches, opportunities, and challenges, IEEE Network 33 (3) (2019) 196–203.

CHAPTER 2

Third-order tensor-based cardiac disease detection from 12-lead ECG signals using deep convolutional neural network

Chhaviraj Chauhan[a], Rajesh Kumar Tripathy[b], and Monika Agrawal[c]

[a]Bharti School of Telecommunication Technology and Management, IIT Delhi, New Delhi, India
[b]Department of EEE, Birla Institute of Technology and Science, Pilani, Hyderabad, India
[c]Centre of Applied Research in Electronics, IIT Delhi, New Delhi, India

Contents

2.1 Introduction

In recent years, the incidence of cardiovascular diseases (CVDs) has been increasing due to unhealthy lifestyles and poor environmental conditions. CVDs, or heart diseases, are fatal for humans and are the leading cause of death among all diseases [1]. Typically, in these diseases, the heart fails to supply an adequate amount of oxygenated blood to other parts of the body, and normal heart function is impaired, resulting in heart muscle damage caused by a lack of adequate oxygen [2]. Early and accurate diagnosis of these conditions is critical in preventing further damage to heart muscles and saving the patient's life. A 12-lead electrocardiogram (ECG) test is the primary and least expensive method of determining whether a person has CVDs [3]. ECG is a noninvasive way of recording the bioelectric signal of the heart's activity. Each heartbeat in the ECG signal consists of a

P-wave, QRS-complex, and T-wave. Cardiologists diagnose cardiac abnormalities by examining the morphological and temporal waves. The amount of recorded ECG data is enormous; diagnosing CVDs manually using all the recorded data is difficult and not practical. As a result, an algorithm for automatic 12-lead ECG analysis is required.

Several approaches have been proposed for automatically detecting CVDs using single- or 12-lead ECG recordings. Two paradigms are used to evaluate these algorithms: intrapatient and interpatient. The intrapatient scheme requires the classifier to learn new parameters using the first few beats or frames of each new patient in the training set [4]. A doctor must manually label the ECG signal of a new patient, which introduces bias into the evaluation process. In the interpatient scheme, the classifier uses the entire beats of certain patients for training and new patient beats for testing. This approach avoids the overlap between the training and testing datasets [5]. As a result, methods based on interpatient paradigms are the most suitable for real-world cardiac diagnostic applications.

Several methods for detecting cardiac diseases are found in the literature based on the intrapatient paradigm [6–10] and the interpatient paradigm [11–13]. These methods for cardiac diseases included in the literature survey were evaluated using a variety of machine learning and deep learning techniques such as KNN [6], CNN [7,13], and LSTM [9,11]. Any method for detecting cardiac diseases is dependent upon the effectiveness of data preprocessing and feature extraction techniques. The feature extraction step is critical for multivariate signals such as 12-lead ECG signals because it must preserve various interlead, intralead, and intrabeat correlations. Recent methods obtain a variety of features, including multiscale phase alternation detection [6] and deep learning-based automated feature extraction [7] [11]. Most existing methods utilize either single-lead ECG signals or univariate techniques on 12-lead ECG signals, which fail to exploit inter- and intralead correlation. Thus, a multivariate technique is required for automatic cardiac disease detection to obtain more discriminating features.

The purpose of this work is to propose an intra- and interpatient paradigm-based cardiac disease detection method with high classification accuracy without sacrificing information about interlead, intralead, and intrabeat correlations. A multivariate projection-based fixed boundary empirical wavelet transform (MPFBEWT) was introduced in [14] to decompose multichannel EEG signals into modes or components. In this method, the empirical wavelet filter bank is designed using fixed boundary points instead of the local maxima- or minima-based boundary detection in the Fourier spectrum of the signal. The boundaries in the frequency spectrum are predefined in this method, which makes the algorithm lightweight and computationally efficient. The MPFBEWT algorithm was also modified by introducing the optimum direction cosine (ODC) concept and renamed MPFBEWT-ODC [14]. The novelty of the proposed work is to form a 3D tensor consisting of samples, leads, and modes after the decomposition of the multilead ECG signals using MPFBEWT-ODC, followed by the use of a suitable deep learning-based classification model for automatic feature extraction and classification of various CVDs. The contribution of this work is listed as follows:

1. The MPFBEWT is explored to decompose the 12-lead ECG beats into the IMFs.
2. The ODC is introduced for the projection in MPFBEWT.
3. A third-order tensor (lead × mode × sample) is formulated for each beat of the 12-lead ECG.
4. A deep CNN-based classifier is used for the detection of cardiac diseases.

The rest of this chapter is divided into the following sections. In Section 2.2, the dataset used is described. The proposed method for detecting heart diseases is discussed in Section 2.3. Section 2.4 presents the results along with a discussion. The conclusion of this paper is presented in Section 2.5.

2.2 Dataset description

The performance of the proposed method is initially evaluated on an open source dataset from the Physikalisch Technische Bundesanstalt (PTB) diagnostic ECG database [15], [16]. In this work, we have taken 12-lead ECG data of 16 myocardial infarction (MI), 20 heart muscle disease (HMD), 16 branch bundle block (BBB), and 16 healthy control (HC) cases from the PTB database. Each signal in this dataset is sampled at a 1000 Hz sampling rate. The second dataset used in this work is from the China Physiological Signal Challenge (CPSC) 2018 database [17], which consists of 6877 12-lead ECG recordings of sampled at 500 Hz. This dataset consists of 12-lead ECG recordings of nine categories: normal, atrial fibrillation (AF), first-degree atrioventricular block (I-AVB), left BBB (LBBB), right BBB (RBBB), premature atrial contraction (PAC), premature ventricular contraction (PVC), ST-segment depression (STD), and ST-segment elevation

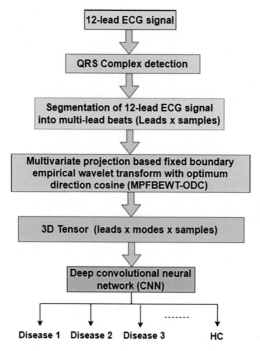

Figure 2.1 A pictorial view of the proposed method for cardiac disease detection.

(STE). In this work, we have used 100 12-lead ECG recordings of each category of cardiac disease from the second database.

2.3 Proposed method

A block diagram of the proposed approach for cardiac disease detection is shown in Fig. 2.1. The proposed method mainly consists of four steps. The first step is prepossessing of the ECG signal and QRS detection. The second step is beat segmentation for multilead ECG signals. The third step is the use of the MPFBEWT–ODC method to decompose each multilead beat. The final step is CNN classifier detection of cardiac diseases. The specific steps are outlined and discussed below.

2.3.1 Preprocessing and beat segmentation

The ECG recording obtained by attaching electrodes to the skin is con-taminated by artifacts due to factors like muscle contraction and breathing

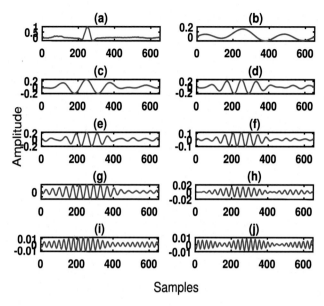

Figure 2.2 Decomposition of the ECG beat into nine modes. (a) Single-lead beat. (b–j) Modes after decomposition of the ECG beat by MPFBEWT.

noise. Therefore, the initial step is to eliminate artifacts by preprocessing the ECG recordings with N leads. In this work, N-lead ($N = 12$) ECG recordings are filtered leadwise by a high-pass filter with a 0.5 Hz cutoff frequency and a low-pass filter with a 100 Hz cutoff frequency. The QRS-complex temporal location in each lead is detected by a single-lead QRS detector [18]. For a beat or cardiac cycle, if the QRS-complex is detected in at least half of the leads, then a common and accurate QRS-complex location is detected by the multilead fusion (MLF) method [19] for each multilead beat. Once the QRS-complex location is accurately detected, the N-lead ECG recordings are segmented into multilead beats. For each detected R-peak of the QRS-complex, 250 ms before the R-peak and 400 ms after the R-peak are considered for the segmentation [20]. A complete beat consisting of a P-wave, QRS-complex, and T-wave is captured in this interval. All these multilead beats are further decomposed into various modes by the MPFBEWT-ODC method as shown in Fig. 2.2 and discussed below.

2.3.2 Multivariate projection-based fixed boundary empirical wavelet transform (MPFBEWT)

The multivariate EWT is an extension of EWT for the analysis of multichannel signals [21]. The objective of EWT is to segment the spectrum of a signal and construct a corresponding wavelet filter bank. In this work, an MPFBEWT filter bank [14] for the decomposition of multilead ECG is used. In MPFBEWT, the N-channel multivariate signal is projected on the unit vector whose direction cosines are equal. Therefore, the direction cosine for each element of the unit vector is considered as $\pm\frac{1}{\sqrt{N}}$. The unit direction vector is given as

$$\hat{i} = \pm\frac{1}{\sqrt{N}}\hat{i}_1 \pm \frac{1}{\sqrt{N}}\hat{i}_2 \pm \ldots \pm \frac{1}{\sqrt{N}}\hat{i}_N. \tag{2.1}$$

The projection of an N-lead ECG signal is calculated as [14]

$$Pr_N = \pm\frac{1}{\sqrt{N}}b_{1p}[n]\hat{i}_1 \pm \frac{1}{\sqrt{N}}b_{2p}[n]\hat{i}_2 \pm \ldots \pm \frac{1}{\sqrt{N}}b_{Np}[n]\hat{i}_N, \tag{2.2}$$

where $b_{1p}[n], b_{2p}[n], \ldots, b_{Lp}[n]$ are the pth beatwise segmented ECG signals corresponding to the respective lead and $\hat{i}_1, \hat{i}_2, \ldots, \hat{i}_N$ are the unit vectors of N channels or leads. The direction cosine for projection is not unique for all multivariate biological signals. Therefore, these should be chosen carefully. To select a set of ODCs, we have considered the morphological features of ECG signal, such as the QRS-complex normally pointing downward in aVR lead and R-wave progression in precordial leads. Based on the aforementioned ECG signal's morphology, an experiment is conducted for various combinations, and a set of ODCs is selected as $-\frac{1}{\sqrt{N}}$ for the fourth, sixth, seventh, and eighth leads (aVR, aVF, V1, V2) and $+\frac{1}{\sqrt{N}}$ for the remaining leads. The projected signal obtained in (2.2) is used to construct an MPFBEWT filter bank for multilead signals. The boundary points in the spectrum of the signal can be found in two ways:
1. adaptive boundary by finding local maxima in the spectrum,
2. fixed boundary in the spectrum.

In the case of nonstationary signals like ECG, the first approach may lead to improper segmentation or boundary detection due to the generation of unwanted local maxima by noise or nonstationary components [22]. Therefore, the fixed boundary points are considered to design the filter bank. These boundary points are computed from the prior initialized frequency points. Most of the useful information of the ECG signal lies be-

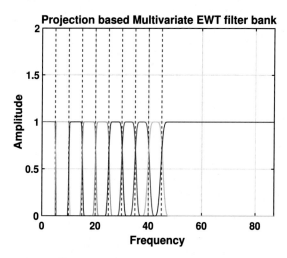

Figure 2.3 MPFBEWT filter bank obtained from the spectrum of a projected ECG signal.

low the frequency of 45 Hz. Therefore, we have divided the spectrum into the frequency ranges of 0–5 Hz, 5–10 Hz, 10–15 Hz, 15–20 Hz, 20–25 Hz, 25–30 Hz, 30–35 Hz, 35–40 Hz, and 40–45 Hz, as shown in Fig. 2.3. In this work, the frequency points $F = [5, 10, 15, 20, 25, 30, 35, 40, 45]$ are used to design the empirical wavelet filter bank. The ith boundary point BP_i can be calculated by the ith frequency point as

$$BP_i = \frac{2\pi \times F_i}{F_s},\tag{2.3}$$

where F_s is the sampling frequency. After obtaining boundary points, the frequency spectrum of the projected signal in the range $[0, \pi]$ is partitioned into M segments [21]. Each segment SG_i is defined as $SG_i = [BP_{i-1}, BP_i]$, where $i = 1, 2, 3...M$, and the concatenation of all boundary points $\bigcup_{i=1}^{M} SG_i = [0, \pi]$ should cover the entire Fourier spectrum of the projected signal. An EWT filter bank is created by empirical scaling and the wavelet function using the segments computed from the Fourier domain of the projected ECG signal [14]. Mathematically, the empirical scaling function ($\hat{\alpha}_i$) and wavelet function are given as

$$
\hat{\alpha}_i(\omega) = \begin{cases} 1, & \text{if } |\omega| \leq BP_i - \xi_i, \\ \cos\left[\frac{\pi}{2}\gamma\left(\frac{1}{2\xi_i}\left(|\omega| - BP_i + \xi_i\right)\right)\right], \\ & \text{if } BP_i - \xi_i \leq |\omega| \leq BP_i + \xi_i, \\ 0, & \text{otherwise} \end{cases} \tag{2.4}
$$

and

$$
\hat{\Phi}_i(\omega) = \begin{cases} 1, & \text{if } BP_i + \xi_i \leq |\omega| \leq BP_{i+1} - \xi_{i+1}, \\ \cos\left[\frac{\pi}{2}\gamma\left(\frac{1}{2\xi_{i+1}}\left(|\omega| - BP_{i+1} + \xi_{i+1}\right)\right)\right], \\ & \text{if } BP_{i+1} - \xi_{i+1} \leq |\omega| \leq BP_{i+1} + \xi_{i+1}, \\ \sin\left[\frac{\pi}{2}\gamma\left(\frac{1}{2\xi_i}\left(|\omega| - BP_i + \xi_i\right)\right)\right], \\ & \text{if } BP_i - \xi_i \leq |\omega| \leq BP_i + \xi_i, \\ 0, & \text{otherwise,} \end{cases} \tag{2.5}
$$

where ξ_i is half of the transition phase width at the ith boundary point and the factor $\gamma(q)$ is given as $\gamma(q) = 35q^4 - 84q^5 + 70q^6 - 20q^7$ [21]. The empirical wavelet filter bank is used to obtain the subband signals or modes in each lead of the ECG signal. The MPFBEWT is applied beatwise on each ECG signal, which yields a 3D tensor of the size (mode × samples × lead). In this work, nine modes are extracted from a multilead beat of a 12-lead ECG signal having 651 samples in each beat. Therefore, 3D tensors of size (9 × 651 × 12) are constructed for each multilead beat. This 3D tensor is fed to a deep CNN for classification.

2.3.3 Deep convolutional neural network (CNN)

A deep CNN [20] is generally used for the effective extraction of spatiotemporal features from raw inputs like videos and 12-lead ECG [13]. The convolutional networks are composed of successive convolutional (filtering) and pooling layers followed by one or more fully connected (FC) layers. Table 2.1 shows the detailed architecture of the CNNs used in this study. The 3D tensor of size (mode × samples × lead (channel)) obtained by the MPFBEWT-ODC is used as the input for the deep CNN. As shown in Table 2.1, three convolutional layers are used. The output feature map $M_{j,k}^l$ of the lth convolutional layer is obtained as [23]

$$
M_{j,k}^l = g\left[\sum_{m=0}^{u-1}\sum_{n=0}^{v-1}\mathcal{K}_{mn}^l M_{(j+m)(k+n)}^{l-1} + b^l\right], \tag{2.6}
$$

Table 2.1 CNN structure.

Layer type	Number of neurons in the layer	Size of kernel	Stride
Input	$9 \times 651 \times 12$	–	–
Convolutional	$9 \times 651 \times 16$	$3 \times 3 \times 16$	[1 1]
Batch normalization	–	–	–
Max-pooling	$4 \times 25 \times 16$	2×2	[2 2]
Convolutional	$4 \times 325 \times 32$	$3 \times 3 \times 32$	[1 1]
Batch normalization	–	–	–
Max-pooling	$2 \times 162 \times 32$	2×2	[2 2]
Convolutional	$2 \times 162 \times 64$	$3 \times 3 \times 64$	[1 1]
Batch normalization	–	–	–
Dropout layer	–	–	–
Full connected	$1 \times 1 \times 4$	–	–
Softmax	$1 \times 1 \times 4$	–	–

where $\mathcal{K}^l_{mn}$ represents the kernel that evaluates the feature map $M^l_{j,k}$ from the previous layer feature map $M^{l-1}_{j,k}$, the size of the kernel is denoted as u and v, b^l is the bias, and $g[\bullet]$ represents the ReLU activation function. Each convolutional layer is followed by the batch normalization (BN) layer to accelerate the training and to improve the training stability [23]. The BN layer normalizes the feature map $M^l_{j,k}$ as follows [23]:

$$B_{j,k} = \frac{M^l_{j,k} - \mu_m}{\sigma_m}, \qquad (2.7)$$

where $B_{j,k}$, μ_m, and σ_m are the normalized value, mean of the batch, and standard deviation within the batch, respectively. After the BN layer, the feature map of the pooling layer for l^{th} is obtained as [23]

$$P^l_{j,k} = max\left[M^{l-1}_{(j+m)(k+n)} \right], \qquad (2.8)$$

where the operator $max[\bullet]$ gives the maximum value of the elements in the input feature map. In this work, three blocks of convolutional, BN, and pooling layers are used. After that, a dropout layer with 50% dropout is used to avoid overfitting. Finally, an FC layer with four neurons followed by a softmax layer is used as the output layer. The output of each neuron with a softmax function is evaluated as [23]

$$S(i) = \frac{e^{w_i \mathbf{u}}}{\sum_{i=1}^{I} e^{w_i \mathbf{u}}}, \tag{2.9}$$

where w_i and $\mathbf{u}$ denote the weight vector of the output layer's ith neuron and the feature vector of the last FC layer, respectively. This CNN model is trained using the stochastic gradient descent with momentum (SGDM) optimizer with a batch size of 32. The loss function for the classification tasks is categorical cross-entropy. One padding for the convolution layer and zero padding for the max-pooling layer have been applied to preserve the output shapes.

2.4 Results and discussion

In this section, the results of the proposed cardiac disease detection method in terms of sensitivity (Se), specificity (Sp), and average classification accuracy (Acc) are shown [24]. Table 2.2 shows the performance of the method for various combinations of direction cosines of the MPFBEWT technique. It has been found that the ODC for leads aVR, aVF, V1, and V2 is $-\frac{1}{\sqrt{N}}$, and it is $+\frac{1}{\sqrt{N}}$ for rest of the leads. The MPFBEWT with ODC provides the best performance, with a sensitivity of 97.37%, a specificity of 99.01%, and an accuracy of 98.53%. It has been demonstrated that the direction cosine according to the morphology of an ECG signal gives the best performance for a projection-based multivariate technique. In Table 2.3, the performance of the cardiac disease detection method is shown for various multivariate techniques such as fast and adaptive multivariate empirical mode decomposition (FA-MVEMD) [25], multivariate variational mode decomposition (MVMD) [26], MPFBEWT [14], and MPFBEWT-ODC. It can be observed that the MPFBEWT-ODC has the highest sensitivity, specificity, and average classification accuracy. The MPFBEWT-ODC-based proposed method is implemented using MATLAB® software (2020b) installed on a workstation (Xeon W-2195 processor, 18 cores, 128 GB RAM, and a processor speed of 2.30 GHz). This automatic cardiac disease detection method was evaluated using K-fold ($K = 5, 10$) cross-validation using intra- and interpatient paradigms. The obtained results are shown in Table 2.4. It is observed that the highest average classification accuracy of the proposed method is obtained by 5-fold cross-validation for the PTB dataset. Similarly, for the CPSC 2018 dataset, a higher accuracy value is observed using 10-fold cross-validation. For the intrapatient paradigm, beats of all ECG recordings are randomly divided for the training and validation sets for K-fold cross-validation. In contrast, for interpatient paradigm-based model evaluation, all beats of a patient are considered as a unit to ensure an interpatient paradigm so that all beats of a recording should lie in one set, i.e., either the training set or the validation set of K-fold cross-validation. In this way, the model will be validated on a different person's ECG recording than the training set's ECG recording. The interpatient paradigm-based method is more realistic and challenging because the model trained on the data of a particular patient may fail on a new patient due to differences in medical history and age. Therefore, the classification accuracy of interpatient paradigm-based methods is always lower than that of intrapatient paradigm-based methods.

Table 2.2 Optimum direction cosine for the proposed method by the intrapatient paradigm.

Direction cosine (sign of $\frac{1}{\sqrt{N}}$) with 12 leads												Se (%)	Sp (%)	Acc (%)
I	II	III	aVR	aVL	aVF	V1	V2	V3	V4	V5	V6			
+	+	+	+	+	+	+	+	+	+	+	+	96.79	98.82	98.28
+	+	+	−	+	−	−	−	+	+	+	+	97.37	99.01	98.53
+	+	+	−	+	−	−	−	+	+	+	+	96.56	98.75	98.18
+	+	+	+	+	+	−	−	−	+	+	+	96.70	98.82	98.31
+	+	+	−	+	+	−	−	−	+	+	+	96.73	98.80	98.26
+	+	+	+	+	−	−	−	−	+	+	+	96.79	98.82	98.28
+	+	+	−	+	−	−	+	+	+	+	+	96.57	98.74	98.17

Table 2.3 Comparison of different multivariate techniques for cardiac disease detection on the PTB dataset with the intrapatient scheme.

Multivariate technique	Se (%)	Sp (%)	Acc (%)
MPFBEWT-ODC	97.37	99.01	98.53
MPFBEWT	96.79	98.82	98.28
FAMVEMD	96.37	98.67	98.06
MVMD	96.56	98.75	98.17

The proposed method is initially evaluated on the PTB dataset. Therefore, the dimensions of the data at every step are shown according to the PTB dataset. The CPSC 2018 dataset is used as the additional test dataset to verify the generalization of the model. The results for the cardiac disease detection method in terms of standard performance criteria [13], including

Table 2.4 Performance of proposed method with different validations.

Dataset	Method type	Validation	Se (%)	Sp (%)	Acc (%)
PTB	Intra	5-fold	97.37	99.01	98.53
		10-fold	96.85	98.85	98.32
	Inter	5 fold	71.60	90.11	85.36
		10-fold	69.70	89.61	84.52
CPSC 2018	Intra	5-fold	97.84	99.72	99.53
		10-fold	97.99	99.74	99.53
	Inter	5-fold	56.48	94.46	90.20
		10-fold	56.60	94.46	90.21

sensitivity, specificity, and classification accuracy, are shown in Table 2.4 and Table 2.5. In Table 2.4, it is observed that the highest average classification accuracy of the proposed method is obtained by 5-fold cross-validation for the PTB dataset and by 10-fold cross-validation for the CPSC 2018 dataset. Fig. 2.4 shows the performance improvement of the proposed method by using the MLF method over the Pan–Tompkins (PT) QRS detection method. Initially, the PT QRS detector was used for beat detection in lead I of ECG recordings from the PTB dataset, which obtained a maximum classification accuracy of 98.36% and 80.45% for the intra- and interpatient paradigms, respectively. After that, the detection performance of the Pan–Tompkins QRS detector was improved by the MLF method [19], which obtained improved results with a maximum average classification accuracy of 98.53% and 85.36% for the intra- and interpatient paradigms, respectively.

The additional dataset used in this study, CPSC 2018, has an imbalanced class problem, as it contains a maximum of 1695 RBBB recordings and a minimum of 202 STE recordings. In addition, each recording lasts between 6 and 60 seconds. The model trained with this dataset may be susceptible to overfitting and poor generalization. Therefore, the research reported in [27] utilized a random undersample with a majority–to–minority class ratio of 2:1. In [1], 962 recordings of six classes with a duration of 10 s are taken for validation. Moreover, the work presented by [28] uses only five classes for evaluation. Therefore, the proposed method is evaluated on the 900 recordings of nine classes from the CPSC 2018 dataset. Table 2.5 presents a comparison of the proposed cardiac disease detection method with reported methods on the PTB and CPSC 2018 datasets using both paradigms. Our method has better classification accuracy, sensitivity, and specificity as compared to reported methods. The proposed method yields a

Table 2.5 Comparison of the proposed method with reported methods for the detection of cardiac diseases on two datasets.

Dataset	Reference	Classifier	Classes	Acc (%)	Method type
PTB	[6] 2016	Fuzzy KNN	4	86.09	Intrapatient
	[7] 2019	Deep CNN	7	98.24	
	[8] 2021	GRU	5	98.50	
	Proposed	CNN	4	98.53	
	[11] 2021	LSTM	2	86.59	Interpatient
	Proposed	CNN	4	85.36	
CPSC 2018	[9] 2021	LSTM	9	94.05	Intrapatient
	[10] 2021	HeartNetEC	9	97.74	
	Proposed	CNN	9	99.53	
	[13] 2021	Deform-CNN	9	86.30	Interpatient
	Proposed	CNN	9	90.21	

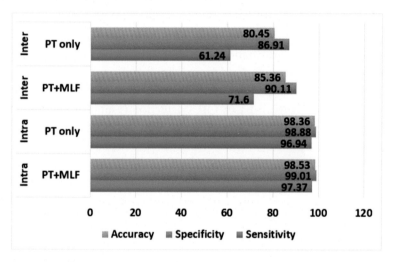

Figure 2.4 Performance metrics of the proposed method on the PTB dataset with the Pan–Tompkins (PT) method and the MLF method for both paradigms.

high classification accuracy, as the more discriminating features are obtained by using a multivariate signal decomposition technique called multivariate projection-based fixed boundary empirical wavelet transform. Also, the ODC for the projection is used according to the morphology of the ECG signal. The boundaries in the frequency spectrum are predefined in this method, which makes the decomposition process lightweight and computationally efficient. In this work, the deep CNN model has been used for the detection of cardiac diseases using the multiscale representation of 12-lead ECG signals. The RNN [5] and transformer models [29] can be used in the multiscale domain of 12-lead ECG for the detection of cardiac diseases.

2.5 Conclusion and summary

This chapter proposes a multiclass classification of cardiac diseases based on intra- and interpatient paradigms using multivariate techniques and deep CNN. The MLF method-based accurate QRS complex detection method has been utilized to achieve a high classification accuracy. The MPFBEWT-ODC has been proposed to decompose the beatwise segmented 12-lead ECG signal in multiple modes to capture interlead, intralead, and intrabeat information. The 3D tensors were formed corresponding to each multi-lead beat, which was used as the input of deep CNN. The deep CNN classified input into four classes for the PTB dataset and nine classes for the CPSC 2018 dataset with high classification accuracy. The key outcomes and motivation for the next problem are as follows:

- The MLF method has been used to detect the QRS-complex. It has been proven that MLF-based QRS detection is superior to single-lead QRS-complex detection in terms of sensitivity, specificity, and classification accuracy.
- The idea of ODC has been introduced for the first time in MPFBEWT for the computation of composite signals using the projection method. Experiments have investigated numerous combinations of the direction cosine to determine the ODC. Due to the unique morphology of multichannel ECG signals, the ODC idea can only be applied to these signals.
- The multiscale decomposition of multilead beat (lead × sample) into modes by MPFBEWT-ODC yields a 3D tensor of size (lead × mode × sample). This 3D tensor contains the information of a beat or cardiac cycle only. However, patient-specific models, which contain the complete information of the patient, are required. Therefore, a higher-order tensor formulation is required.
- For the detection of cardiac diseases, a deep CNN-based classifier is utilized in this work. Deep CNN performed well in terms of accuracy when the input was a 3D tensor. However, as the order of the tensor increases, so does the complexity of the CNN model due to the

convolutional layers. Consequently, a more appropriate classifier with fewer parameters is required.

References

[1] S. Ran, X. Yang, M. Liu, Y. Zhang, C. Cheng, H. Zhu, Y. Yuan, Homecare-oriented ECG diagnosis with large-scale deep neural network for continuous monitoring on embedded devices, IEEE Transactions on Instrumentation and Measurement 71 (2022) 1–13.

[2] L. Lu, M. Liu, R. Sun, Y. Zheng, P. Zhang, Myocardial infarction: symptoms and treatments, Cell Biochemistry and Biophysics 72 (3) (2015) 865–867.

[3] B.J. Maron, R.A. Friedman, P. Kligfield, B.D. Levine, S. Viskin, B.R. Chaitman, P.M. Okin, J.P. Saul, L. Salberg, G.F. Van Hare, et al., Assessment of the 12-lead ECG as a screening test for detection of cardiovascular disease in healthy general populations of young people (12–25 years of age): a scientific statement from the American Heart Association and the American College of Cardiology, Circulation 130 (15) (2014) 1303–1334.

[4] G. De Lannoy, D. François, J. Delbeke, M. Verleysen, Weighted conditional random fields for supervised interpatient heartbeat classification, IEEE Transactions on Biomedical Engineering 59 (1) (2011) 241–247.

[5] W. Liu, F. Wang, Q. Huang, S. Chang, H. Wang, J. He, MFB-CBRNN: a hybrid network for MI detection using 12-lead ECGs, IEEE Journal of Biomedical and Health Informatics 24 (2) (2019) 503–514.

[6] R. Tripathy, S. Dandapat, Detection of cardiac abnormalities from multilead ECG using multiscale phase alternation features, Journal of Medical Systems 40 (6) (2016) 143.

[7] N.I. Hasan, A. Bhattacharjee, Deep learning approach to cardiovascular disease classification employing modified ECG signal from empirical mode decomposition, Biomedical Signal Processing and Control 52 (2019) 128–140.

[8] A. Darmawahyuni, S. Nurmaini, M.N. Rachmatullah, F. Firdaus, B. Tutuko, Unidirectional-bidirectional recurrent networks for cardiac disorders classification, Telkomnika 19 (3) (2021) 902–910.

[9] C.-Y. Chen, Y.-T. Lin, S.-J. Lee, W.-C. Tsai, T.-C. Huang, Y.-H. Liu, M.-C. Cheng, C.-Y. Dai, Automated ECG classification based on 1D deep learning network, Methods 202 (2022) 127–135, https://doi.org/10.1016/j.ymeth.2021.04.021.

[10] S.A. Deevi, C.P. Kaniraja, V.D. Mani, D. Mishra, S. Ummar, C. Satheesh, Heart-NetEC: a deep representation learning approach for ECG beat classification, Biomedical Engineering Letters 11 (1) (2021) 69–84.

[11] A. Rath, D. Mishra, G. Panda, LSTM-based cardiovascular disease detection using ECG signal, in: Cognitive Informatics and Soft Computing, Springer, 2021, pp. 133–142.

[12] R. Ge, T. Shen, Y. Zhou, C. Liu, L. Zhang, B. Yang, Y. Yan, J.-L. Coatrieux, Y. Chen, Convolutional squeeze-and-excitation network for ECG arrhythmia detection, Artificial Intelligence in Medicine 121 (2021) 102181.

[13] L. Qin, Y. Xie, X. Liu, X. Yuan, H. Wang, An end-to-end 12-leading electrocardiogram diagnosis system based on deformable convolutional neural network with good antinoise ability, IEEE Transactions on Instrumentation and Measurement 70 (2021) 1–13.

[14] R.K. Tripathy, S.K. Ghosh, P. Gajbhiye, U.R. Acharya, Development of automated sleep stage classification system using multivariate projection-based fixed boundary empirical wavelet transform and entropy features extracted from multichannel EEG signals, Entropy 22 (10) (2020) 1141.

[15] A.L. Goldberger, L.A. Amaral, L. Glass, J.M. Hausdorff, P.C. Ivanov, R.G. Mark, J.E. Mietus, G.B. Moody, C.-K. Peng, H.E. Stanley, PhysioBank, PhysioToolkit, and PhysioNet: components of a new research resource for complex physiologic signals, Circulation 101 (23) (2000) e215–e220.

[16] M. Oeff, H. Koch, R. Bousseljot, D. Kreiseler, The PTB Diagnostic ECG Database, National Metrology Institute of Germany, 2012.

[17] F. Liu, C. Liu, L. Zhao, X. Zhang, X. Wu, X. Xu, Y. Liu, C. Ma, S. Wei, Z. He, et al., An open access database for evaluating the algorithms of electrocardiogram rhythm and morphology abnormality detection, Journal of Medical Imaging and Health Informatics 8 (7) (2018) 1368–1373.

[18] J. Pan, W.J. Tompkins, A real-time QRS detection algorithm, IEEE Transactions on Biomedical Engineering ME-32 (3) (1985) 230–236.

[19] Chhaviraj Chauhan, Monika Agrawal, Pooja Sabherwal, Accurate QRS complex detection in 12-lead ECG signals using multi-lead fusion, Measurement 223 (2023) 113776.

[20] U.R. Acharya, H. Fujita, S.L. Oh, Y. Hagiwara, J.H. Tan, M. Adam, Application of deep convolutional neural network for automated detection of myocardial infarction using ECG signals, Information Sciences 415 (2017) 190–198.

[21] J. Gilles, Empirical wavelet transform, IEEE Transactions on Signal Processing 61 (16) (2013) 3999–4010.

[22] Y. Hu, F. Li, H. Li, C. Liu, An enhanced empirical wavelet transform for noisy and non-stationary signal processing, Digital Signal Processing 60 (2017) 220–229.

[23] N. Muralidharan, S. Gupta, M.R. Prusty, R.K. Tripathy, Detection of COVID19 from X-ray images using multiscale deep convolutional neural network, Applied Soft Computing 119 (2022) 108610.

[24] M. Sokolova, G. Lapalme, A systematic analysis of performance measures for classification tasks, Information Processing & Management 45 (4) (2009) 427–437.

[25] M.R. Thirumalaisamy, P.J. Ansell, Fast and adaptive empirical mode decomposition for multidimensional, multivariate signals, IEEE Signal Processing Letters 25 (10) (2018) 1550–1554.

[26] N. ur Rehman, H. Aftab, Multivariate variational mode decomposition, IEEE Transactions on Signal Processing 67 (23) (2019) 6039–6052.

[27] Z. Li, H. Zhang, Automatic detection for multi-labeled cardiac arrhythmia based on frame blocking preprocessing and residual networks, Frontiers in Cardiovascular Medicine 8 (2021) 616585.

[28] C. Ma, K. Lan, J. Wang, Z. Yang, Z. Zhang, Arrhythmia detection based on multiscale fusion of hybrid deep models from single lead ECG recordings: a multicenter dataset study, Biomedical Signal Processing and Control 77 (2022) 103753.

[29] A. Vaswani, N. Shazeer, N. Parmar, J. Uszkoreit, L. Jones, A.N. Gomez, Ł. Kaiser, I. Polosukhin, Attention is all you need, Advances in Neural Information Processing Systems 30 (2017).

CHAPTER 3

Ramanujan filter bank-domain deep CNN for detection of atrial fibrillation using 12-lead ECG

Abhay Patwari[a], Shaswati Dash[a], Rajesh Kumar Tripathy[a], Ganapati Panda[b], and Ram Bilas Pachori[c]

[a]Department of EEE, Birla Institute of Technology and Science, Pilani, Hyderabad, India
[b]Department of ECE, CV Raman Global University, Bhubaneswar, India
[c]Department of Electrical Engineering, Indian Institute of Technology Indore, Indore, India

Contents

3.1 Introduction

Atrial fibrillation (AF) results in an abnormal and irregular heart rhythm that leads to blood clots in the heart and increases the risk of myocardial in-farction (MI), heart failure, and sudden cardiac arrest [1]. In AF, the heart's atria beat chaotically, and there is a lack of synchronization between the atrial and ventricular activities. AF episodes lasting less than 24 hours are interpreted as paroxysmal AF (PAF), whereas in persistent AF, the abnormal atrial activity lasts more than seven days [2]. In a clinical study, the 12-lead ECG test is performed to diagnose AF [3]. Pathological alterations such as atrial fibrillatory (f)-waves rather than the normal P-wave episode and variations in the interbeat duration (RR interval) are used to detect AF-based cardiac arrhythmia [4]. The advances in wearable technologies enable the continuous monitoring of 12-lead ECG signals for diagnosing AF [5]. The constant recording of 12-lead ECG generates a considerable volume of data. Hence, it is a cumbersome process for medical staff to manually examine each ECG beat to detect AF pathology. Therefore, artificial intelligence (AI)-based methods are required to automatically detect AF using 12-lead ECG signals [6]. The AI-based techniques for automated detection of AF are based on either machine learning (extraction of features from ECG signal using different signal processing methods and use of different classifiers to detect AF) or deep learning algorithms [2]. Deep learning models have the advantage of automatically performing learnable feature extraction and classification as the manual feature-engineering-based techniques for detecting AF using ECG time series. Developing novel deep learning-based techniques to automatically detect AF using 12-lead ECG recording is important in cardiovascular engineering.

In recent years, different 12-lead ECG-based deep learning approaches have been proposed for the automated detection of AF [7]. Back et al. [8] used deep bidirectional long-short-term memory (BiLSTM) to detect AF using the 12-lead ECG signal. Similarly, Cai et al. [3] considered a 36-layer-based deep neural network and 12-lead ECG data for detecting AF. Ribeiro et al. [9] used the deep residual neural network model with input 12-lead ECG signals to detect MI. In another work, Biton et al. [10] have considered deep representation learning to predict AF risk using 12-lead ECG signals. Similarly, other deep learning techniques such as LSTM [11] and transformer [12] models have been used to classify arrhythmia using ECG signals. The methods mentioned earlier to detect AF directly consider the 12-lead ECG signals as input to the deep learning models. Joint time-frequency transform-domain-based deep learning models have been used to analyze ECG and other biomedical signals [2] [13] [14]. For AF-based cardiac disease detection using single-lead ECG signals, the time-frequency-based deep learning models have shown higher classification performance than the time-domain ECG signal-based deep neural networks [2]. The input joint time-frequency-domain physiological signal information helps the deep neural network to extract robust discriminative features for the automated detection of different ailments.

The Ramanujan filter bank (RFB) is a signal processing method used to evaluate and track the periodicity of the time series [15]. This filter bank has been explored for detecting the QRS-complex in ECG signals [16]. In AF-based cardiac ailments, the chaotic atrial f-waves and the normal QRS-complex with variations in the RR intervals are observed in the ECG signal [4]. The RFB-based time-period representation (TPR) can effectively capture the periodicity in the QRS-complex and f-waves in the

ECG signals during AF. In addition, the deep learning model developed using the TPR of ECG signals can be used to detect AF. The RFB–domain deep CNN architecture has not been employed to detect AF-based cardiac ailments using ECG signals. The novelty of the proposed work is to develop a TPR-domain multichannel deep CNN architecture to automatically detect AF using 12-lead ECG recordings. The significant contributions of the proposed work are highlighted as follows:

- RFB is used to obtain the TPRs of 12-lead ECG signals for normal sinus rhythm (NSR) and AF classes.
- A deep CNN architecture is proposed in the RFB domain of 12-lead ECG signals for optimal lead selection.
- The TPRs of selected ECG leads and a novel multichannel deep CNN are used to detect AF.
- The results obtained using the RFB-based deep learning approach are compared with other deep learning- and time-frequency-based methods to detect AF.

The remaining parts or sections of this chapter are organized as follows. The 12-lead ECG dataset and the proposed AF detection approach are dealt with in Sections 3.2 and 3.3, respectively. The results evaluated using the proposed approach and the analysis of these results are shown in Section 3.4. We have outlined the conclusions of this chapter in Section 3.5.

3.2 12-lead ECG database

This study uses a publicly available 12-lead ECG dataset (from the Shaoxing database) [17] to develop and assess the proposed RFB-based deep learning approach. This dataset contains 12-lead ECG recordings of 10,646 subjects. Out of these 10,646 subjects, 5956 are males and 4690 are females, aged between 4 and 98 years. The sampling frequency of 12-lead ECG data for the Shaoxing dataset is 500 Hz, and the signals were recorded using the GE MUSE ECG system [17]. In this dataset, out of 10,646 recordings, 1826 and 1780 12-lead ECG recordings belong to the NSR and AF classes, respectively. The length of each 12-lead ECG signal is 10 seconds. We have considered all 12-lead ECG recordings from both AF and NSR classes to develop and evaluate the performance of the RFB-based deep learning techniques.

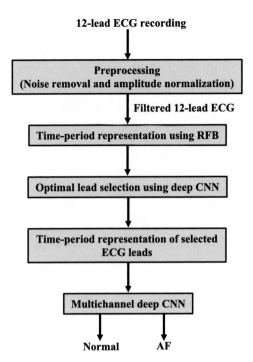

Figure 3.1 Block diagram representation of the proposed AF detection methodology using 12-lead ECG signals.

3.3 Proposed approach

A block diagram of the 12-lead ECG-based approach to detect AF is shown in Fig. 3.1. It consists of the preprocessing of 12-lead ECG data, the evaluation of the TPR of each lead ECG frame using RFB, and a deep CNN for AF detection. The preprocessing of 12-lead ECG data involves the elimination of baseline wandering noise using an IIR filter (Butterworth filter with order two) with a cutoff frequency of 0.5 Hz and normalizing the amplitude value of the filtered ECG signal [2]. Amplitude normalization is performed by dividing the amplitude of each channel ECG signal by the maximum value. After preprocessing, the TPR of each channel ECG signal is processed using RFB.

3.3.1 Time-period representation of ECG

The Ramanujan sum is defined based on the mathematical expression as $c_m(n) = \sum_{k=1}^{m} e^{2\pi j \frac{k}{m} n}$ [18], where the greatest common divisor (GCD) of k

and m is unity. Here, k and m are coprime to each other. The Ramanujan sum is periodic, $c_m(n + m) = c_m(n)$, and $-\infty \leqslant n \leqslant \infty$ [16]. The $c_m(n)$ can also be computed [15] as

$$c_m(n) = \mu \left(\frac{m}{\text{GCD}(m, n)} \right) \frac{\Phi(m)}{\Phi \left(\frac{m}{\text{GCD}(m,n)} \right)}, \tag{3.1}$$

where $\Phi(m)$ is interpreted as the Euler totient function and $\mu(\bullet)$ is called the Mobius function. The Ramanujan sum is a real sum, and $c_m(n)$ is an integer for each value of n. In RFB, $c_m(n)$ denotes the impulse response of the filter [16]. The mathematical expression for the frequency response $C_m(e^{jw})$ computed based on the impulse response $c_m(n)$ is given as

$$C_m(e^{jw}) = 2\pi \sum_{k=1}^{m} \delta \left(w - \frac{2\pi k}{m} \right). \tag{3.2}$$

For a signal $z(n)$ with T as the period, the discrete Fourier transform is given as

$$Z(e^{jw}) = \frac{2\pi}{T} \sum_{r=1}^{T-1} Z(r)\delta \left(w - \frac{2\pi r}{T} \right). \tag{3.3}$$

By comparing (3.2) and (3.3), it is observed that if the signal $z(n)$ is used as input to the Ramanujan sum-based impulse response in (3.2), then the factor $C_m(e^{jw})$ is nonzero when the period T is a multiple of m. The set of the filters from the Ramanujan sum $C_m(n)$ within the period T with $1 \leq m \leq N$ is interpreted as RFB. The period T is an important parameter for obtaining the TPR using RFB [16]. In this study, the value of T is selected as 250 for RFB-based analysis of ECG signals. The lead I ECG signals for NSR and AF cases are illustrated in Fig. 3.2(a) and Fig. 3.2(c), respectively. The TPRs of NSR- and AF-based ECG signals are shown in Fig. 3.2(b) and Fig. 3.2(d), respectively. In the TPR of the AF ECG signal, high-intensity patterns appear for higher period values. The TPR characteristics show a significant difference for AF and NSR classes. Therefore, the TPR images can be considered as the input to the deep CNN to detect AF automatically.

3.3.2 Development of TPR-domain deep CNN

In this study, the TPR-domain deep CNN [19] is developed to detect AF. The deep CNN model constructed using the TPR of a single-lead EEG

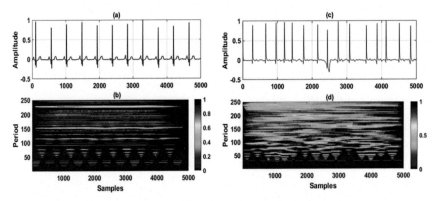

Figure 3.2 (a) ECG signal plot for NSR class. (b) TPR of NSR ECG signal evaluated using RFB. (c) ECG signal plot for AF class. (b) TPR of AF ECG signal evaluated using RFB.

signal is depicted in Fig. 3.3. This CNN model consists of four 2D convolution layers, two 2D max-pooling layers, a flattened layer, and an output layer. The performance of this deep CNN architecture is evaluated using TPR images of each lead ECG signal. Based on the classification accuracy of the single-channel deep CNN model, the ECG signals of 3 out of 12 leads are selected. Similarly, the architecture of the multichannel deep CNN with as input the TPR images of the three best ECG leads is depicted in Fig. 3.4. The multichannel deep CNN comprises 11 2D convolution layers, 4 2D max-pooling layers, a flattened layer, and an output layer, respectively. For each deep CNN model, the hold-out (70% training, 10% validation, and 20% testing) and 5-fold cross-validation (CV)-based methodologies are employed for selecting training and testing 12-lead ECG instances [13]. Hyperparameters such as the "Adam" optimizer, a learning rate of 0.01, a batch size of 256, and 100 epochs are considered for the single-channel deep CNN model of Fig. 3.3. The binary cross-entropy is used as the loss function during the training phases of both deep CNN models [2]. For the multichannel deep CNN case, as in Fig. 3.4, a learning rate of 0.001 is used. All other hyperparameters of this model are the same as that of the single-channel deep CNN model. The performance metrics such as sensitivity, kappa score, specificity, accuracy, and F1-score are used to assess the performance of both deep CNN models [13] [20].

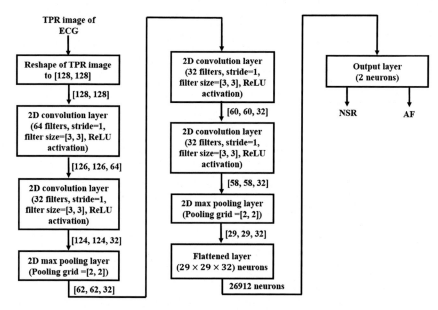

Figure 3.3 Block diagram of deep CNN to automatically detect AF using the TPR image of a single-lead ECG signal.

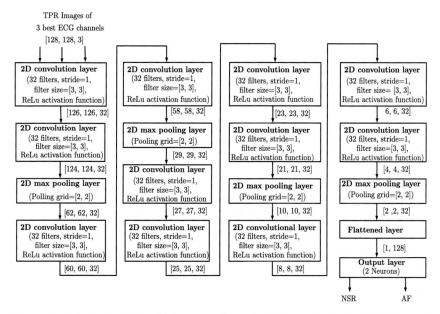

Figure 3.4 A detailed block diagram of a deep CNN model using TPR images of the three best ECG leads for the detection of AF.

Table 3.1 Classification results of hold-out CV.

Lead	Accuracy	Sensitivity	Specificity	Kappa	F1-score
I	95.10	97.51	92.66	0.9017	0.9520
II	96.07	97.73	94.79	0.9210	0.9603
III	92.84	94.48	91.17	0.8563	0.9293
aVR	95.05	95.67	94.42	0.9007	0.9507
aVL	91.63	94.39	88.85	0.8327	0.9187
aVF	94.59	93.84	95.35	0.8913	0.9453
V1	94.96	96.41	93.49	0.8990	0.9503
V2	94.32	98.34	90.25	0.886	0.945
V3	95.01	95.31	94.70	0.900	0.950
V4	91.27	85.12	97.49	0.825	0.907
V5	95.70	98.62	92.75	0.914	0.958
V6	94.45	96.41	92.47	0.889	0.945

3.4 Results and discussion

The classification results obtained using the single-channel deep CNN model with the TPRs of ECG signals of all 12 leads using hold-out CV are shown in Table 3.1. It is noticed that the accuracy and F1-score values of deep CNN are higher than 90% using TPR images of each lead ECG signal. Similarly, except for TPR images of lead III and lead V4 ECG signals, the sensitivity values of deep CNN are higher than 93% using TFR images of other ECG leads. Moreover, except for the TPR images of lead aVL ECG signals, deep CNN has produced specificity values higher than 90% using TPR images of ECG signals of other leads. The 5-fold CV-based results obtained using a deep CNN model with the TPR of each lead ECG signal for AF detection are presented in Fig. 3.5. It is evident that for ECG signals of leads I, II, aVR, aVL, V2, V3, and V6, the accuracy, sensitivity, and specificity values computed using deep CNN are higher than 90% using TPR images. Similarly, using the TPR images of lead aVF, lead V4, and lead V5 ECG signals, the sensitivity value of deep CNN is less than 90%. The three best ECG leads chosen based on the accuracy values of the deep CNN model with hold-out CV are leads I, II, and V5, respectively. Moreover, the classification results of the multichannel deep CNN architecture evaluated using the TPR images of 3-lead ECG signals are presented in Table 3.2. The accuracy of multichannel deep CNN is obtained as 98.14%, which is higher than that of single-channel deep CNN using TPR images evaluated using ECG signals. Moreover, the multichannel deep CNN classification results have been

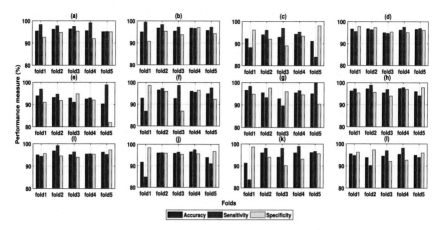

Figure 3.5 Performance measures (accuracy, sensitivity, and specificity) obtained using deep CNN with TPR images of ECG signals for (a) lead I, (b) lead II, (c) lead III, (d) lead aVR, (e) lead aVL, (f) lead aVF, (g) lead V1, (h) lead V2, (i) lead V3, (j) lead V4, (k) lead V5, and (l) lead V6.

Table 3.2 Classification performance of multichannel deep CNN using TPR images from three selected leads.

Lead selection	Accuracy	Sensitivity	Specificity	Kappa	F1-score
(I, II, V5)	98.14	97.40	98.88	0.962	0.980
(I, aVF, V5)	97.88	97.07	98.71	0.957	0.978

obtained using the images of other time-frequency-based representations, namely, the short-time Fourier transform (STFT), chirplet transform, and continuous wavelet transform (CWT), of ECG signals to detect AF. These results are shown in Table 3.3. The multichannel deep CNN architecture with as input STFT-based TFR images of 3-lead ECG signals obtained accuracy, kappa score, sensitivity, and specificity values of 66.62%, 0.330, 87.05%, and 45.96%, respectively. Similarly, the multichannel deep CNN with as input the chirplet transform-based TFRs of 3-lead ECG signals have demonstrated higher accuracy than CWT- and STFT-domain TFRs of 3-lead ECG signals. The RFB-based TPR of 3-lead ECG signals and multichannel deep CNN has achieved superior classification performance for the detection of AF as compared to other TFRs of 3-lead ECG signals.

The comparison of the results of the proposed TPR–domain multichannel deep CNN with those of multilayer LSTM- and transformer-based

Table 3.3 Comparison of different performance metrics of deep CNN with other time-frequency-based methods using three selected ECG leads.

Deep learning methods	Accuracy	Sensitivity	Specificity	Kappa	F1-score
STFT and deep CNN	66.62	87.05	45.96	0.330	0.723
CWT and deep CNN	86.84	80.44	93.31	0.737	0.860
Chirplet transform and deep CNN	88.50	92.56	84.40	0.769	0.890
RFB and deep CNN	98.14	97.40	98.88	0.962	0.980

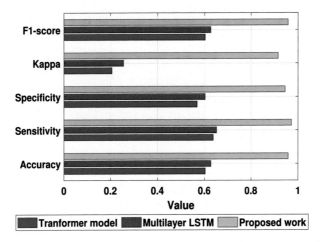

Figure 3.6 Comparison with existing methods for the detection of AF using three selected ECG leads (LSTM- [11] and transformer-based [12] models).

deep learning models is shown in Fig. 3.6. The accuracy, sensitivity, specificity, kappa score, and F1–score of the transformer model are 64.34%, 63.75%, 56.84%, 0.21, and 0.60, respectively. Similarly, for the multilayer LSTM model, the accuracy, sensitivity, specificity, kappa score, and F1–score are 62.79%, 65.21%, 60.30%, 0.26, and 0.63, respectively. It is noteworthy that the proposed TPR-based deep CNN outperformed the multilayer LSTM- and transformer-based models in the automatic detection of AF-based cardiac disease using 3-lead ECG signals. Moreover, the significance of the difference in accuracy values evaluated using deep learning-based methods shown in Fig. 3.6 for AF detection is investigated using analysis of variance (ANOVA) [13]. If $p < 0.001$, this signifies that the difference in all performance measures between the proposed AF detection approach and the existing deep learning methods (transformer and multilayer LSTM) for detecting AF is significant. The advantages of the RFB-based proposed deep learning approach are outlined as follows:

- The TPR images of lead I, II, and V5 ECG signals provide higher classification performance using deep CNN.
- The TPR representation obtained using the RFB of ECG signals potentially capture the pathological changes during AF.
- The RFB-based TPR coupled with multichannel deep CNN yields superior classification performance compared to other TFRs of ECG data.
- The multichannel deep CNN with TPR images of 3-lead ECG signals obtains higher classification accuracy than LSTM and transformer models.

In the current investigation, only the NSR versus AF classification task has been carried out using the TPR images of ECG signals and multichannel deep CNN. In the future, the proposed RFB-based approach can be used for the automated detection of paroxysmal and persistent AF episodes [2] and other cardiac ailments using 12-lead ECG signals.

3.5 Conclusion

A novel approach to automatically detect AF using TPRs of 12-lead ECG signals has been proposed in this work. RFB has been used to obtain the TPR of each lead ECG signal. The deep CNN model 1 architecture has been proposed to select the optimal 3 leads out of the 12 leads of ECG signals. The TPR of the three best lead ECG signals (I, II, and V5) and the multichannel deep CNN architecture have been employed to detect AF. The proposed RFB-based deep CNN approach has attained higher overall classification accuracy than the transformer- and multilayer LSTM-based deep learning techniques. The proposed deep CNN with RFB-based TPR has provided superior performance compared to other TFRs of 3-lead ECG signals. Finally, the RFB-based deep CNN model can detect other cardiac ailments using 12-lead ECG recordings.

References

[1] C.S. Boerschel, R.B. Schnabel, The imminent epidemic of atrial fibrillation and its concomitant diseases – myocardial infarction and heart failure - a cause for concern, International Journal of Cardiology 287 (2019) 162–173.
[2] T. Radhakrishnan, J. Karhade, S.K. Ghosh, P.R. Muduli, R. Tripathy, U.R. Acharya, AFCNNet: automated detection of AF using chirplet transform and deep convolu-

tional bidirectional long short term memory network with ECG signals, Computers in Biology and Medicine 137 (2021) 104783.

[3] W. Cai, Y. Chen, J. Guo, B. Han, Y. Shi, L. Ji, J. Wang, G. Zhang, J. Luo, Accurate detection of atrial fibrillation from 12-lead ECG using deep neural network, Computers in Biology and Medicine 116 (2020) 103378.

[4] S.K. Ghosh, R.K. Tripathy, M.R. Paternina, J.J. Arrieta, A. Zamora-Mendez, G.R. Naik, Detection of atrial fibrillation from single lead ECG signal using multirate cosine filter bank and deep neural network, Journal of Medical Systems 44 (6) (2020) 1–15.

[5] C.-T. Lin, K.-C. Chang, C.-L. Lin, C.-C. Chiang, S.-W. Lu, S.-S. Chang, B.-S. Lin, H.-Y. Liang, R.-J. Chen, Y.-T. Lee, et al., An intelligent telecardiology system using a wearable and wireless ECG to detect atrial fibrillation, IEEE Transactions on Information Technology in Biomedicine 14 (3) (2010) 726–733.

[6] Z. Chen, J. Luo, K. Lin, J. Wu, T. Zhu, X. Xiang, J. Meng, An energy-efficient ECG processor with weak-strong hybrid classifier for arrhythmia detection, IEEE Transactions on Circuits and Systems. II, Express Briefs 65 (7) (2017) 948–952.

[7] F. Murat, F. Sadak, O. Yildirim, M. Talo, E. Murat, M. Karabatak, Y. Demir, R.-S. Tan, U.R. Acharya, Review of deep learning-based atrial fibrillation detection studies, International Journal of Environmental Research and Public Health 18 (21) (2021) 11302.

[8] Y.-S. Baek, S.-C. Lee, W. Choi, D.-H. Kim, A new deep learning algorithm of 12-lead electrocardiogram for identifying atrial fibrillation during sinus rhythm, Scientific Reports 11 (1) (2021) 1–10.

[9] A.H. Ribeiro, M.H. Ribeiro, G.M. Paixão, D.M. Oliveira, P.R. Gomes, J.A. Canazart, M.P. Ferreira, C.R. Andersson, P.W. Macfarlane, W. Meira Jr, et al., Automatic diagnosis of the 12-lead ECG using a deep neural network, Nature Communications 11 (1) (2020) 1–9.

[10] S. Biton, S. Gendelman, A.H. Ribeiro, G. Miana, C. Moreira, A.L.P. Ribeiro, J.A. Behar, Atrial fibrillation risk prediction from the 12-lead electrocardiogram using digital biomarkers and deep representation learning, European Heart Journal-Digital Health 2 (4) (2021) 576–585.

[11] S. Saadatnejad, M. Oveisi, M. Hashemi, LSTM-based ECG classification for continuous monitoring on personal wearable devices, IEEE Journal of Biomedical and Health Informatics 24 (2) (2019) 515–523.

[12] B. Wang, C. Liu, C. Hu, X. Liu, J. Cao, Arrhythmia classification with heartbeat-aware transformer, in: ICASSP 2021-2021 IEEE International Conference on Acoustics, Speech and Signal Processing (ICASSP), IEEE, 2021, pp. 1025–1029.

[13] J. Karhade, S. Dash, S.K. Ghosh, D.K. Dash, R.K. Tripathy, Time–frequency-domain deep learning framework for the automated detection of heart valve disorders using PCG signals, IEEE Transactions on Instrumentation and Measurement 71 (2022) 1–11.

[14] S. Madhavan, R.K. Tripathy, R.B. Pachori, Time-frequency domain deep convolutional neural network for the classification of focal and non-focal EEG signals, IEEE Sensors Journal 20 (6) (2019) 3078–3086.

[15] S.V. Tenneti, P. Vaidyanathan, Ramanujan filter banks for estimation and tracking of periodicities, in: 2015 IEEE International Conference on Acoustics, Speech and Signal Processing (ICASSP), IEEE, 2015, pp. 3851–3855.

[16] S. Mukhopadhyay, S. Krishnan, Robust identification of the QRS-complexes in electrocardiogram signals using Ramanujan filter bank-based periodicity estimation technique, Frontiers in Signal Processing 2 (2022) 921973, https://doi.org/10.3389/frsip.2022.921973.

[17] J. Zheng, J. Zhang, S. Danioko, H. Yao, H. Guo, C. Rakovski, A 12-lead electrocardiogram database for arrhythmia research covering more than 10,000 patients, Scientific Data 7 (1) (2020) 1–8.

[18] S. Ramanujan, On certain trigonometrical sums and their applications in the theory of numbers [Trans. Cambridge Philos. Soc. 22 (13) (1918) 259–276], in: Collected Papers of Srinivasa Ramanujan, 2000, pp. 179–199.

[19] I. Goodfellow, Y. Bengio, A. Courville, Deep Learning, MIT Press, 2016, http://www.deeplearningbook.org.

[20] S. Chen, R. Zheng, T. Wang, T. Jiang, F. Gao, D. Wang, J. Cao, Deterministic learning based WEST syndrome analysis and seizure detection on ECG, IEEE Transactions on Circuits and Systems. II, Express Briefs (2022) 4603–4607.

CHAPTER 4

Detection of atrial fibrillation using photoplethysmography signals: a systemic review

Cheuk To Skylar Chung[a]**, Vellaisamy Roy**[b]**, Gary Tse**[c]**, and Haipeng Liu**[d]

[a]Cardiovascular Analytics Group, PowerHealth Research Institute, Hong Kong, China
[b]School of Science and Technology, Hong Kong Metropolitan University, Hong Kong, China
[c]School of Nursing and Health Studies, Hong Kong Metropolitan University, Hong Kong, China
[d]Centre for Intelligent Healthcare, Coventry University, Coventry, United Kingdom

Contents

4.1 Introduction

Atrial fibrillation (AF) is characterized by overexcitation of the atrium, resulting in dyssynchronous atrial and ventricular contraction [1]. Subsequently, this increases the risks of developing stroke, heart failure, and cognitive decline [2,3]. Common risk factors associated with AF include age, genetic predisposition, and hypertension [4]. The prevalence of AF has drastically increased by 33% over two decades in the beginning of the 21st century, with an estimated 4977 cases per million habitants worldwide [5]. To differentiate AF from various types of arrhythmia, an electrocardiogram

Signal Processing Driven Machine Learning Techniques for
Cardiovascular Data Processing
https://doi.org/10.1016/B978-0-44-314141-6.00009-8

(ECG) is utilized. AF can be identified from ECG recordings by the absence of a discrete P-wave, a narrow QRS-complex, and an irregular ventricular rhythm [6]. Furthermore, the type of AF can also be classified based on the frequency and duration of AF episodes, including paroxysmal AF, persistent AF, long-standing persistent AF, and permanent AF [7]. Common treatment options for AF include ablation therapy, anticoagulants, and other pharmacological agents, where accurate and early detection of AF plays a key role [8].

Although ECG is a useful tool for AF detection, recordings are often misinterpreted as other types of arrhythmia such as sinus tachycardia or supraventricular tachycardia. The high cost and limited portability also compromise the accessibility of ECG. In addition, due to their short duration, ECG recordings are often insufficient to capture nonpermanent or asymptomatic types of AF, especially paroxysmal AF [9]. Interestingly, only 17% of AF cases were identified on the first ECG attempt in a Swedish screening program [10]. As approximately 40% of AF patients are asymptomatic, many patients are underdiagnosed prior to stroke and undertreated in clinical practice [9]. While population screening programs for AF have been proposed, this approach remains heavily debated due to questionable cost-efficacy. It is therefore intuitive to explore the application of alternative low-cost approaches in public health strategies to prevent AF. This garnered interest in novel physiological metrics extracted by wearable sensing techniques such as photoplethysmography (PPG) for AF detection. PPG is an optical technique that utilizes a low-intensity light source (green, red, or infrared) and measures light absorption to determine volumetric variations in microcirculation [11]. It is predicted that PPG-based technology, including wearable devices and contactless radars, will become more pervasive in a diverse range of clinical settings [12]. Current evidence of the application of PPG in cardiology, including nociception [13], blood pressure [14], and pulse rate monitoring, is well understood [13,15]. This may extend to the detection of various cardiovascular diseases [16,17]. The recent advancements in PPG-based technology denote promising potential in AF detection as a fast, affordable, and noninvasive alternative to ECG [18]. However, the current literature on the clinical efficacy of PPG-based methods for AF detection is scarce. Thus, this chapter aims to conduct a systemic review of the contemporary literature surrounding the use of PPG signals in the detection of AF.

4.2 Methods

4.2.1 Search strategy, inclusion and exclusion criteria

This systemic review was conducted in accordance to the Preferred Reporting Items for Systematic Reviews and Meta-Analyses statement (PRISMA) [19]. The PubMed database was searched from inception to 1 September, 2023 with no language restrictions. The following search terms were utilized: "atrial fibrillation" AND "PPG." Outcomes that were used to evaluate the validity of PPG-based methods for AF detection include (1) overall accuracy, (2) sensitivity, (3) specificity, (4) positive predictive value (PPV), and (5) negative predictive value (NPV).

The inclusion criteria were as follows: (1) the study was a randomized control trial on human participants, a prospective or retrospective observational study, or a case-control study; (2) the study presents evidence of using PPG-derived physiological metrics for AF detection; (3) the study presents relevant data of at least one of the outcomes of interest; and (4) patients with AF were analyzed. The exclusion criteria were as follows: (1) the study performs AF detection without PPG signal processing; (2) the study is not accessible electronically; (3) the study is incomplete; (4) the study is an editorial, a poster, or non-peer-reviewed material; and (5) the study is a duplicate of an included publication.

4.2.2 Data extraction

All publications identified from the initial literature search were compiled onto a Microsoft Excel spreadsheet. The full list of studies was later assessed for eligibility by separate reviewers and any discrepancies were discussed between reviewers.

4.3 Results and discussion

For the detection of AF using PPG, a total of 130 articles were retrieved from PubMed, and 18 articles were included for further quantitative analysis. Through a systematic search, we included various real-world studies and randomized control trials targeting patients who underwent screening for AF in various clinical settings. The main findings are as follows: (1) techniques utilizing PPG-derived physiological metrics can accurately detect the presence of AF with high cost-effectiveness and accessibility; (2) long-term monitoring of AF using PPG-based wearable technology can allow for early detection and management; and (3) involvement of ar-

Figure 4.1 Word cloud of the abstracts of included studies.

tificial intelligence (AI) techniques in analyzing PPG data can drastically enhance the efficacy of PPG-based AF detection (Fig. 4.1). Current studies on PPG-based AF detection methods are diverse in device, signal presentation, recording length, classification approach, and number of patients (Fig. 4.2).

4.3.1 Features

Currently, the pulse-to-pulse interval is often extracted from PPG signals for AF detection, although this approach may falsely detect AF in the case of premature atrial contraction or premature ventricular contraction [21]. Under stable measurement conditions, the signal quality of PPG is comparable to that of ECG in detecting interbeat interval features. Eerikäinen et al. demonstrated that the calculated pNN40 (the percentage of successive RR intervals that differ by more than 40 ms) using PPG data appears to be a strong and more robust feature for AF detection compared to the coefficient of sample entropy (CosEn) derived from ECG [22]. Another study highlighted 11 features from PPG data that are relevant for AF detection, which are summarized in Fig. 4.3 [23]. A separate systemic review isolated a total of 44 features from PPG data for AF detection with mixed evidence

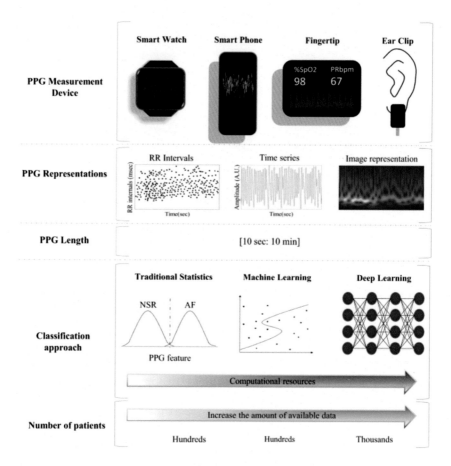

Figure 4.2 Summary of the AF detection process using PPG signals [20].

on individual performance, including 29 analyzed in the time domain, 12 from the frequency domain, and 3 from the time-frequency domain [24].

4.3.2 Cost-effectiveness and accessibility

In comparison to ECG, PPG can offer greater long-term cost–effectiveness and better accessibility for patients. An economic evaluation via a microsimulation model of 30 million individuals reported that the overall preferred screening strategy was wearable PPG, followed by ECG patches across multiple demographic conditions [25]. The Apple Heart study was the first study to suggest that the extraction of pulse irregularities from PPG signals can accurately identify unknown AF in a broad population cohort of

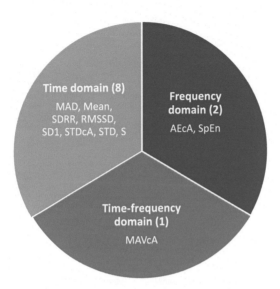

Figure 4.3 Eleven features extracted from PPG signals for AF detection, inspired by Table 3 of Millán et al. [23]. MAD, mean absolute deviation; SDRR, standard deviation of the RR intervals; RMSSD, root mean square of successive differences; SD1, the standard deviation of the Poincaré plot perpendicular to the line of identity; STD, standard deviation; STDcA, standard deviation of approximation coefficients; AEcA, average energy of approximation coefficients; MAVcA, mean absolute value of approximation coefficients; SpEn, spectral entropy.

over 400,000 participants [26]. In the Huawei Heart study, 187,912 Chinese participants were screened for AF using a PPG screening app, in which 262 AF-suspected participants were followed up with full medical examination and entered the Mobile Atrial Fibrillation Application (MAFA) integrated care program. The PPV of this model was reported to be 91.6% [27]. A separate algorithm was designed in the Fitbit Heart study incorporating 455,699 participants, which identified 340 cases of AF from a variety of wrist-worn devices with a PPV of over 97%. Wearing the devices during the night can maximize the sensitivity of the PPG sensors to identify undiagnosed AF [28].

The feasibility of smart technology monitoring is corroborated by examining real-world evidence. Amongst different types of wearable sensors, the most frequently researched device appears to be the smartwatch [29]. In an ambulatory setting, smartwatch-recorded PPG allows for the continuous monitoring of AF and can be considered as an alternative screening modality to standard ECG monitoring [30,31]. Zhu et al. proposed a smart watch PPG-based algorithm which showed only modest differences in mean absolute error for AF detection between daytime and nighttime after four weeks of continuous monitoring [32]. This is especially favorable for elderly patients who are at higher risk of AF but suffer from reduced mobility [33], as demonstrated in a study recruiting 60 participants from a community senior care organization [34].

Interestingly, the heterogeneity observed in smart device monitoring may be attributed to the large-scale disparities in methodology and cohort. Taken together, there is a need for more pragmatic real-world studies to assess the role of smart technology for AF detection. Nevertheless, the use of mobile healthcare technology can provide a substantial enrichment of information for subsequent ECG monitoring and intervention of AF. Overall, the current literature illustrates the benefits of PPG-based continuous home monitoring methods, namely, improving patient comfort and mass screening practicality.

4.3.3 Incorporation of machine and deep learning

Although the concept of AI-assisted disease detection has existed for several decades [35], it was quite recently that AI was introduced into PPG signal processing methods for AF detection. This can be performed through machine learning (ML) and deep learning (DL) models. The former approach is based on feature analysis, while the latter allows PPG data to be inputted directly into the system for automatic extraction of underlying features.

Several studies obtained promising results using random forest (RF) and gradient boosting decision tree (GBDT) models in analyzing PPG and accelerometer data to differentiate AF from other cardiac conditions [36], even allowing for safe monitoring after cardiac surgery [37]. Researchers also designed hybrid models combining 1D convolutional neural network (CNN) and bidirectional long-short term memory network (BiLSTM) to extract both ECG and PPG signals to improve the AF detection accuracy, which performed better compared to a residual network (ResNet) model or individual networks [38]. The lightweight architecture of CNN is important for reducing computational complexity and real-time processing towards real-world practicality. The available evidence suggests that DL approaches exhibit superior performance in PPG-based AF detection relative to ML approaches due to their inherent ability to analyze different types of data and extract underlying deep features, including a range of waveform characteristics [39]. Specifically, Poh et al. found that support vector machine (SVM) had the best performance amongst non-DL techniques, while deep CNN performed better than SVM in detecting AF using pulse waveforms [40]. However, many of these algorithms require further validation using a larger, more varied population cohort under real-world clinical settings. The approaches of common algorithms and AI models are summarized in Table 4.1.

4.3.4 Clinical implications

The 2020 European Society of Cardiology (ESC) guidelines do not recommend to use PPG as a singular screening tool for AF detection [51]. The general consensus of clinicians is that AF identification with ECG is more accurate compared to PPG (83% vs. 27%) [52]. Nonetheless, PPG is already being used in clinical practice for continuous remote monitoring. In fact, the benchmark for presenting PPG data consists of the PPG waveform with its corresponding Poincaré plot and tachogram, which is later reviewed by a physician [45].

This systemic review provides evidence that noninvasive remote monitoring by smart wearable PPG devices can be equally effective to ECG usage which is more expensive and less accessible. Hence, healthcare resources may consider distributing more inpatient care resources for patients who are diagnosed or are at high risk of AF, while implementing more low-cost PPG-based strategies for early detection. As a result, further examination may be directed towards a smaller target population, which can aid in reducing healthcare expenditure. In the future, we may open discussions to reevaluate the currently dominant ECG-based standards in AF detection and consider alternative methods as equally viable (Fig. 4.4).

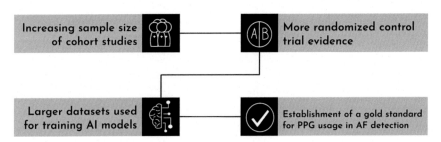

Figure 4.4 Summary of future directions for the development of PPG-based AF detection approaches.

Table 4.1 Diagnostic performance of novel AI algorithms from included trials and cohort studies.

Study	Algorithms of PPG-based AF detection	Sensitivity	Specificity	Accuracy	PPV	NPV
Lubitz et al. [28]	Fitbit Heart algorithm	0.676	0.984	–	0.982	–
Guo et al. [27]	Mobile Atrial Fibrillation Application (MAFA)	1.00	0.990	0.990	0.916	–
Zhu et al. [32]	Hybrid machine learning decision model	0.878	0.974	–	–	–
Eerikäinen et al. [36]	Random forest	0.976	0.982	0.967	0.928	–
Saarinen et al. [31]	PulseOn Arrhythmia Monitor System	0.956	0.992	–	–	–
Perez et al. [26]	Apple Heart smartwatch-based irregular pulse notification algorithm	–	–	–	0.840	–
Aldughayfiq et al. [38]	Hybrid deep neural network (1D CNN and BiLSTM)	0.900	0.850	0.950	–	–
Nemati et al. [41]	Samsung Simband logistics regression	0.970	0.940	0.950	–	–
Ramesh et al. [42]	1D deep CNN	0.946	0.952	0.951	–	–
Liao et al. [43]	Support vector machine and random forest	0.960	0.925	0.952	0.956	0.947
Mutke et al. [44]	Heartbeats algorithm (nonlinear rhythm analysis)	0.954	0.916	0.933	–	–
Gruwez et al. [45]	Combination of FibriCheck algorithm, PPG waveform, tachogram and Poincaré plot	0.975	0.950	0.959	0.555	0.998
Kwon et al. [46]	CNN	0.990	0.943	0.969	0.956	0.987
Sun et al. [47]	3D CNN	0.954	0.911	0.960	–	–
Selder et al. [34]	FibriCheck AF algorithm	1.00	0.96	0.970	0.750	1.00
Väliaho et al. [48]	AFEvidence algorithm	0.962	0.981	–	–	–
Bacevicius et al. [49]	Automated PPG-based algorithm	0.942	0.969	0.968	–	–
Ding et al. [50]	Log-spectral matching GAN (LSM-GAN)	0.928	0.988	0.961	0.985	0.943

Abbreviations: AI, artificial intelligence; PPV, positive predictive value; NPV, negative predictive value; CNN, convolution neural network; BiLSTM, bidirectional long short-term memory.

4.3.5 Limitations and research gaps

While PPG-based approaches are appealing alternatives for scalability and cost-effectiveness, the applicability of PPG-based AF detection could be limited by several factors. The 2023 wearable PPG roadmap summarized the future directions of PPG technology in medicine into five main goals: (1) expanding the functionality of wearable devices; (2) optimizing the sensor design for data collection; (3) exploring the advantages and disadvantages of existing PPG processing approaches; (4) recognizing areas of clinical application; and (5) garnering trust from the public and related stakeholders [53].

From a technical perspective, PPG waveform is susceptible to noises such as environmental light and motion artifacts during daily healthcare monitoring. Unlike the P-wave in ECG, PPG does not have an explicit feature that is indicative of atrial contraction; therefore, AF detection is dependent on finding irregularities in peak-to-peak intervals. A shared concern in the current literature underscores the effect of movement, which may limit the potential of AF detection using smart devices [54]. Resultantly, improper preprocessing can deform the waveform and shift the key features [55,56]. For instance, filtering-induced waveform deformation can cause inaccuracy in evaluating heart rate variability [57]. To mitigate this limitation, many state-of-the-art techniques aim to eliminate motion noise and separating interference signals in PPG processing [41,58,59]. This notwithstanding, further research should be conducted to establish a standardized framework for PPG signal preprocessing [60]. In contrast, ECG signals are stable, and some advanced methods including tailored discrete wave transformation have been developed to effectively preserve the ECG waveform features, including the J-point, i.e., the junction between QRS termination and ST-segment onset [61].

On a physiological level, as PPG waveform reflects the microcirculation in peripheral arteries, the electrophysiological activities of the heart are not reflected directly. It is important to consider that PPG signals can be influenced by other physiological factors, including respiration pattern [62], blood pressure [63], and neural activity [64]. Therefore, the accurate separation of cardiac electrophysiological activities from PPG signals calls for further development in advanced signal processing.

In analyzing PPG-based AF detection approaches, there is a lack of standardization and large-scale validation of existing studies. As of now, studies primarily target algorithmic development and perform validation on small cohorts. Considering the physiological diversity and the existence of comorbidity among different cohorts, the reliability of PPG-based AF detection needs large-scale validation.

4.4 Conclusion

In conclusion, the findings described in this chapter indicate that the performance of PPG-based techniques in AF detection is moderate to high. While some studies suggest that the attained accuracy and sensitivity of these algorithms are equally effective to traditional ECG-based detection, some real-world studies suggest that ECG methods are overall more effective. AI-assisted PPG signal processing is likely to develop in the future with more prospective clinical trials and real-world studies. Early AF detection is imperative in reducing the risk of stroke and improving patient prognosis. The wearable PPG devices can be further integrated into other healthcare settings, optimizing current diagnostic standards of AF. More importantly, it is crucial to reexamine our current clinical guidelines and regulatory frameworks regarding the clinical implementation of PPG-based AF detection.

References

[1] L. Staerk, J.A. Sherer, D. Ko, E.J. Benjamin, R.H. Helm, Atrial fibrillation: epidemiology, pathophysiology, and clinical outcomes, Circulation Research 120 (9) (2017) 1501–1517.

[2] M. Rienstra, S.A. Lubitz, S. Mahida, J.W. Magnani, J.D. Fontes, M.F. Sinner, et al., Symptoms and functional status of patients with atrial fibrillation: state of the art and future research opportunities, Circulation 125 (23) (2012) 2933–2943.

[3] A. Alonso, A.P. Arenas de Larriva, Atrial fibrillation, cognitive decline and dementia, European Cardiology 11 (1) (2016) 49–53.

[4] M.K. Chung, L.L. Eckhardt, L.Y. Chen, H.M. Ahmed, R. Gopinathannair, J.A. Joglar, et al., Lifestyle and risk factor modification for reduction of atrial fibrillation: a scientific statement from the American Heart Association, Circulation 141 (16) (2020) e750–e772.

[5] G. Lippi, F. Sanchis-Gomar, G. Cervellin, Global epidemiology of atrial fibrillation: an increasing epidemic and public health challenge, International Journal of Stroke 16 (2) (2021) 217–221.

[6] S. Biton, S. Gendelman, A.H. Ribeiro, G. Miana, C. Moreira, A.L.P. Ribeiro, et al., Atrial fibrillation risk prediction from the 12-lead electrocardiogram using digital biomarkers and deep representation learning, European Heart Journal - Digital Health 2 (4) (2021) 576–585.

[7] G. Batra, B. Svennblad, C. Held, T. Jernberg, P. Johanson, L. Wallentin, et al., All types of atrial fibrillation in the setting of myocardial infarction are associated with impaired outcome, Heart 102 (12) (2016) 926–933.

[8] C. Gutierrez, D.G. Blanchard, Diagnosis and treatment of atrial fibrillation, American Family Physician/GP 94 (6) (2016) 442–452.

[9] V. Thijs, Atrial fibrillation detection: fishing for an irregular heartbeat before and after stroke, Stroke 48 (10) (2017) 2671–2677.

[10] E. Svennberg, J. Engdahl, F. Al-Khalili, L. Friberg, V. Frykman, M. Rosenqvist, Mass screening for untreated atrial fibrillation: the STROKESTOP study, Circulation 131 (25) (2015) 2176–2184.

[11] J. Park, H.S. Seok, S.S. Kim, H. Shin, Photoplethysmogram analysis and applications: an integrative review, Frontiers in Physiology 12 (2021) 808451.

[12] R.S. Vulcan, S. André, M. Bruyneel, Photoplethysmography in normal and pathological sleep, Sensors (Basel) 21 (9) (2021).

[13] M. Koeny, Analyzing pain and stress from PPG perfusion signal patterns, in: Studies in Skin Perfusion Dynamics: Photoplethysmography and Its Applications in Medical Diagnostics, 2021, pp. 151–161.

[14] X. Xing, Z. Ma, M. Zhang, Y. Zhou, W. Dong, M. Song, An unobtrusive and calibration-free blood pressure estimation method using photoplethysmography and biometrics, Scientific Reports 9 (1) (2019) 8611.

[15] J.A. van der Stam, E.H.J. Mestrom, J. Scheerhoorn, F. Jacobs, S. Nienhuijs, A.K. Boer, et al., The accuracy of wrist-worn photoplethysmogram-measured heart and respiratory rates in abdominal surgery patients: observational prospective clinical validation study, JMIR Perioperative Medicine 6 (2023) e40474.

[16] T. Sadad, S.A.C. Bukhari, A. Munir, A. Ghani, A.M. El-Sherbeeny, H.T. Rauf, Detection of cardiovascular disease based on PPG signals using machine learning with cloud computing, Computational Intelligence and Neuroscience 2022 (2022) 1672677.

[17] D. Castaneda, A. Esparza, M. Ghamari, C. Soltanpur, H. Nazeran, A review on wearable photoplethysmography sensors and their potential future applications in health care, International Journal of Biosensors & Bioelectronics 4 (4) (2018) 195–202.

[18] M.A. Almarshad, M.S. Islam, S. Al-Ahmadi, A.S. BaHammam, Diagnostic features and potential applications of PPG signal in healthcare: a systematic review, Healthcare (Basel) 10 (3) (2022) 11–111.

[19] D. Moher, L. Shamseer, M. Clarke, D. Ghersi, A. Liberati, M. Petticrew, et al., Preferred reporting items for systematic review and meta-analysis protocols (PRISMA-P) 2015 statement, Systematic Reviews 4 (1) (2015) 1–9.

[20] T. Pereira, N. Tran, K. Gadhoumi, M.M. Pelter, D.H. Do, R.J. Lee, et al., Photoplethysmography based atrial fibrillation detection: a review, npj Digital Medicine 3 (1) (2020) 3.

[21] D. Han, S.K. Bashar, F. Zieneddin, E. Ding, C. Whitcomb, D.D. McManus, et al., Digital image processing features of smartwatch photoplethysmography for cardiac arrhythmia detection, Annual International Conference of the IEEE Engineering in Medicine and Biology Society 2020 (2020) 4071–4074.

[22] L.M. Eerikäinen, A.G. Bonomi, F. Schipper, L.R.C. Dekker, R. Vullings, H.M. de Morree, et al., Comparison between electrocardiogram- and photoplethysmogram-derived features for atrial fibrillation detection in free-living conditions, Physiological Measurement 39 (8) (2018) 084001.

[23] C.A. Millán, N.A. Girón, D.M. Lopez, Analysis of relevant features from photoplethysmographic signals for atrial fibrillation classification, International Journal of Environmental Research and Public Health 17 (2) (2020).

[24] N.A. Girón, C.A. Millán, D.M. López, Systematic review on features extracted from PPG signals for the detection of atrial fibrillation, Studies in Health Technology and Informatics 261 (2019) 266–273.

[25] W. Chen, S. Khurshid, D.E. Singer, S.J. Atlas, J.M. Ashburner, P.T. Ellinor, et al., Cost-effectiveness of screening for atrial fibrillation using wearable devices, JAMA Health Forum 3 (8) (2022) e222419.

[26] M.V. Perez, K.W. Mahaffey, H. Hedlin, J.S. Rumsfeld, A. Garcia, T. Ferris, et al., Large-scale assessment of a smartwatch to identify atrial fibrillation, The New England Journal of Medicine 381 (20) (2019) 1909–1917.

[27] Y. Guo, H. Wang, H. Zhang, T. Liu, Z. Liang, Y. Xia, et al., Mobile photoplethys-mographic technology to detect atrial fibrillation, Journal of the American College of Cardiology 74 (19) (2019) 2365–2375.

[28] S.A. Lubitz, A.Z. Faranesh, C. Selvaggi, S.J. Atlas, D.D. McManus, D.E. Singer, et al., Detection of atrial fibrillation in a large population using wearable devices: the Fitbit heart study, Circulation 146 (19) (2022) 1415–1424.

[29] A. Sijerčić, E. Tahirović, Photoplethysmography-based smart devices for detection of atrial fibrillation, The Texas Heart Institute Journal 49 (5) (2022).

[30] P.-C. Chang, M.-S. Wen, C.-C. Chou, C.-C. Wang, K.-C. Hung, Atrial fibrillation detection using ambulatory smartwatch photoplethysmography and validation with simultaneous Holter recording, American Heart Journal 247 (2022) 55–62.

[31] H.J. Saarinen, A. Joutsen, K. Korpi, T. Halkola, M. Nurmi, J. Hernesniemi, et al., Wrist-worn device combining PPG and ECG can be reliably used for atrial fibrillation detection in an outpatient setting, Frontiers in Cardiovascular Medicine 10 (2023) 1100127.

[32] L. Zhu, V. Nathan, J. Kuang, J. Kim, R. Avram, J. Olgin, et al., Atrial fibrilla-tion detection and atrial fibrillation burden estimation via wearables, IEEE Journal of Biomedical and Health Informatics 26 (5) (2022) 2063–2074.

[33] F. Babar, A.M. Cheema, Z. Ahmad, A. Sarfraz, Z. Sarfraz, H. Ashraff, et al., Sensitivity and specificity of wearables for atrial fibrillation in elderly populations: a systematic review, Current Cardiology Reports 25 (7) (2023) 761–779.

[34] J. Selder, T. Proesmans, L. Breukel, O. Dur, W. Gielen, A.C. van Rossum, et al., Assessment of a standalone photoplethysmography (PPG) algorithm for detection of atrial fibrillation on wristband-derived data, Computer Methods and Programs in Biomedicine 197 (2020) 105753.

[35] V. Kaul, S. Enslin, S.A. Gross, History of artificial intelligence in medicine, Gastroin-testinal Endoscopy 92 (4) (2020) 807–812.

[36] L.M. Eerikainen, A.G. Bonomi, F. Schipper, L.R.C. Dekker, H.M. de Morree, R. Vullings, et al., Detecting atrial fibrillation and atrial flutter in daily life using pho-toplethysmography data, IEEE Journal of Biomedical and Health Informatics 24 (6) (2020) 1610–1618.

[37] D. Hiraoka, T. Inui, E. Kawakami, M. Oya, A. Tsuji, K. Honma, et al., Diagnosis of atrial fibrillation using machine learning with wearable devices after cardiac surgery: algorithm development study, JMIR Formative Research 6 (8) (2022) e35396.

[38] B. Aldughayfiq, F. Ashfaq, N.Z. Jhanjhi, M. Humayun, A deep learning approach for atrial fibrillation classification using multi-feature time series data from ECG and PPG, Diagnostics (Basel) 13 (14) (2023).

[39] S. Kwon, J. Hong, E.K. Choi, E. Lee, D.E. Hostallero, W.J. Kang, et al., Deep learning approaches to detect atrial fibrillation using photoplethysmographic signals: algorithms development study, JMIR mHealth and uHealth 7 (6) (2019) e12770.

[40] M.-Z. Poh, Y.C. Poh, P.-H. Chan, C.-K. Wong, L. Pun, W.W.-C. Leung, et al., Diagnostic assessment of a deep learning system for detecting atrial fibrillation in pulse waveforms, Heart 104 (23) (2018) 1921–1928.

[41] S. Nemati, M.M. Ghassemi, V. Ambai, N. Isakadze, O. Levantsevych, A. Shah, et al., Monitoring and detecting atrial fibrillation using wearable technology, Annual International Conference of the IEEE Engineering in Medicine and Biology Society 2016 (2016) 3394–3397.

[42] J. Ramesh, Z. Solatidehkordi, R. Aburukba, A. Sagahyroon, Atrial fibrillation classi-fication with smart wearables using short-term heart rate variability and deep convo-lutional neural networks, Sensors (Basel) 21 (21) (2021).

[43] M.T. Liao, C.C. Yu, L.Y. Lin, K.H. Pan, T.H. Tsai, Y.C. Wu, et al., Impact of recording length and other arrhythmias on atrial fibrillation detection from wrist pho-toplethysmogram using smartwatches, Scientific Reports 12 (1) (2022) 5364.

[44] M.R. Mutke, N. Brasier, C. Raichle, F. Ravanelli, M. Doerr, J. Eckstein, Comparison and combination of single-lead ECG and photoplethysmography algorithms for wearable-based atrial fibrillation screening, Telemedicine Journal and E-Health 27 (3) (2021) 296–302.

[45] H. Gruwez, S. Evens, T. Proesmans, D. Duncker, D. Linz, H. Heidbuchel, et al., Accuracy of physicians interpreting photoplethysmography and electrocardiography tracings to detect atrial fibrillation: INTERPRET-AF, Frontiers in Cardiovascular Medicine 8 (2021) 734737.

[46] S. Kwon, J. Hong, E.K. Choi, B. Lee, C. Baik, E. Lee, et al., Detection of atrial fibrillation using a ring-type wearable device (CardioTracker) and deep learning analysis of photoplethysmography signals: prospective observational proof-of-concept study, Journal of Medical Internet Research 22 (5) (2020) e16443.

[47] Z. Sun, J. Junttila, M. Tulppo, T. Seppänen, X. Li, Non-contact atrial fibrillation detection from face videos by learning systolic peaks, IEEE Journal of Biomedical and Health Informatics 26 (9) (2022) 4587–4598.

[48] E-S. Väliaho, P. Kuoppa, J.A. Lipponen, T.J. Martikainen, H. Jäntti, T.T. Rissanen, et al., Wrist band photoplethysmography in detection of individual pulses in atrial fibrillation and algorithm-based detection of atrial fibrillation, Europace 21 (7) (2019) 1031–1038.

[49] J. Bacevicius, Z. Abramikas, E. Dvinelis, D. Audzijoniene, M. Petrylaite, J. Marinskiene, et al., High specificity wearable device with photoplethysmography and six-lead electrocardiography for atrial fibrillation detection challenged by frequent premature contractions: DoubleCheck-AF, Frontiers in Cardiovascular Medicine 9 (2022) 869730.

[50] C. Ding, R. Xiao, D.H. Do, D.S. Lee, R.J. Lee, S. Kalantarian, et al., Log-spectral matching GAN: PPG-based atrial fibrillation detection can be enhanced by GAN-based data augmentation with integration of spectral loss, IEEE Journal of Biomedical and Health Informatics 27 (3) (2023) 1331–1341.

[51] G. Hindricks, T. Potpara, N. Dagres, E. Arbelo, J.J. Bax, C. Blomström-Lundqvist, et al., 2020 ESC Guidelines for the diagnosis and management of atrial fibrillation developed in collaboration with the European Association for Cardio-Thoracic Surgery (EACTS) The Task Force for the diagnosis and management of atrial fibrillation of the European Society of Cardiology (ESC) Developed with the special contribution of the European Heart Rhythm Association (EHRA) of the ESC, European Heart Journal 42 (5) (2021) 373–498.

[52] M. Manninger, D. Zweiker, E. Svennberg, S. Chatzikyriakou, N. Pavlovic, J.A. Zaman, et al., Current perspectives on wearable rhythm recordings for clinical decision-making: the wEHRAbles 2 survey, Europace 23 (7) (2021) 1106–1113.

[53] P.H. Charlton, J. Allen, R. Bailon, S. Baker, J.A. Behar, F. Chen, et al., The 2023 wearable photoplethysmography roadmap, Physiological Measurement 44 (11) (2023) 111001.

[54] S. Ismail, U. Akram, I. Siddiqi, Heart rate tracking in photoplethysmography signals affected by motion artifacts: a review, EURASIP Journal on Advances in Signal Processing 2021 (1) (2021) 5.

[55] H. Liu, J. Allen, S.G. Khalid, F. Chen, D. Zheng, Filtering-induced time shifts in photoplethysmography pulse features measured at different body sites: the importance of filter definition and standardization, Physiological Measurement 42 (7) (2021) 074001.

[56] S. Liao, H. Liu, W.-H. Lin, D. Zheng, F. Chen, Filtering-induced changes of pulse transmit time across different ages: a neglected concern in photoplethysmography-based cuffless blood pressure measurement, Frontiers in Physiology 14 (2023).

[57] E. Mejía-Mejía, J.M. May, P.A. Kyriacou, Effect of filtering of photoplethysmography signals in pulse rate variability analysis, in: 2021 43rd Annual International Conference of the IEEE Engineering in Medicine & Biology Society (EMBC), IEEE, 2021.

[58] S.K. Bashar, D. Han, S. Hajeb-Mohammadalipour, E. Ding, C. Whitcomb, D.D. Mc-Manus, et al., Atrial fibrillation detection from wrist photoplethysmography signals using smartwatches, Scientific Reports 9 (1) (2019) 15054.

[59] A.H.A. Zargari, S.A.H. Aqajari, H. Khodabandeh, A.M. Rahmani, F. Kurdahi, An accurate non-accelerometer-based PPG motion artifact removal technique using Cy-cleGAN, arXiv preprint, arXiv:2106.11512, 2021.

[60] P.H. Charlton, K. Pilt, P.A. Kyriacou, Establishing best practices in photoplethys-mography signal acquisition and processing, Physiological Measurement 43 (5) (2022) 050301.

[61] X. Zhao, J. Zhang, Y. Gong, L. Xu, H. Liu, S. Wei, et al., Reliable detection of myocardial ischemia using machine learning based on temporal-spatial characteristics of electrocardiogram and vectorcardiogram, Frontiers in Physiology 13 (2022) 854191.

[62] H. Liu, F. Chen, V. Hartmann, S.G. Khalid, S. Hughes, D. Zheng, Comparison of different modulations of photoplethysmography in extracting respiratory rate: from a physiological perspective, Physiological Measurement 41 (9) (2020) 094001.

[63] S.G. Khalid, H. Liu, T. Zia, J. Zhang, F. Chen, D. Zheng, Cuffless blood pressure esti-mation using single channel photoplethysmography: a two-step method, IEEE Access 8 (2020) 58146–58154.

[64] S.G. Khalid, S.M. Ali, H. Liu, A.G. Qurashi, U. Ali, Photoplethysmography tempo-ral marker-based machine learning classifier for anesthesia drug detection, Medical & Biological Engineering & Computing 60 (11) (2022) 3057–3068.

CHAPTER 5

Machine learning-based prediction of depression and anxiety using ECG signals

Ramnivas Sharma and Hemant Kumar Meena
Malaviya National Institute of Technology (MNIT), Jaipur, Rajasthan, India

Contents

5.1 Introduction

Stress, anxiety, and depression (SAD) are psychological illnesses that seriously affect mental stability, disturb someone's regular daily activities, and deteriorate into trauma in extreme situations. When a person is stressed, sad, or worried, their body releases different chemicals. This changes their nonverbal body language. Due to the many stages of their research, these mental illnesses can be grouped as stress, anxiety, and depression [1]. Stress is the stage of mental illness. During this stage, anxiety keeps the mind from becoming unstable [2]. Anxiety and depression is the most severe stage of a mental illness that can have a long-term detrimental effect on a person's physical and mental health [3]. The amount of stress a person feels directly affects how uncomfortable they feel, and this discomfort shows up as anxiousness or sadness. Exercises, extra work, too many tasks, heavy breathing, not getting enough sleep, questionnaires, and other things can all cause stress. According to one study [4], stress can have positive and negative effects depending on the situation. Recognizing stress and looking at how it shows up in different ways to diagnose a person is a complicated process. Medical science makes it difficult for computer vision [5] techniques to diagnose or treat stress; however, SAD can be effectively diagnosed and treated using computerized computer vision techniques. These artificial intelligence (AI)-based solutions have a competitive advantage over medical investigations because their systems work quickly and on their own [6]. The process of stress recognition is made simpler and more suitable by machine learning approaches because of the availability of numerous datasets. The patient's voice, video, ECG and EEG signals, skin conductance, skin temperature, blood pressure, and other data are all included in the dataset. These datasets [7] give important information about how and why a patient's body works and reacts, which can be used to study stress recognition. The next critical phase is to obtain features from the provided data. The dataset includes significant characteristics that demonstrate the remarkable diversity of psychological activity. This chapter examines the research on the discovery and exploration of stress levels using machine learning approaches. The study demonstrates how well various models identify psychiatric conditions linked to stress, anxiety, and depression.

Anxiety is the most common type of mental illness, with numerous effects on the person and society. Physical symptoms are common in people with anxiety disorders, and as a result, many of these patients consult their primary care physicians. Mental diseases, generalized anxiety disorder (GAD), disorders with excessive fear, and associated behavioral problems are included in the category of anxiety disorders [8]. Anxiety is recognized as a very complicated emotional integrity component that characterizes the individual. In Europe, stress is the most common mental illness. It affects 16% of the population [9]. One of the factors that are frequently mentioned is stress, and much research has been done on its diagnosis and treatment. Microelectronics, sensor technology, machine learning, and data networks have become increasingly important since the development of computing technologies. Electroencephalogram [10] and electrocardiogram [11] signal acquisition, wearable body sensors [12], and data mining techniques [13] are just a few of the options suggested in the literature. Anxiety disorder diagnosis is a complicated and tough undertaking. As a result, one must take care to diagnose them accurately. The patient's history can be looked at to determine what is wrong, and machine learning and data mining methods can mimic how humans think or come to logical conclusions. Some techniques can even operate on incomplete or questionable data using notions from pattern matching, probability, and other domains [14].

5.2 Mental health problems

A mental health issue, also known as a mental disease, is a condition that affects how someone feels, thinks, behaves, and interacts with others [15]. The American Psychiatric Association defines mental health issues as emotional, cognitive, behavioral, or a combination of these conditions linked to social, occupational, or familial functioning issues. A mental health condition interferes with one's ability to think, feel, act, or communicate with others in society. Therefore, problems with mental health can also affect daily tasks. Anxiety, depression, stress, and schizophrenia disorders are mental health issues [16]. In Malaysia, depression is the most prevalent mental health issue, followed by stress and anxiety [17].

5.2.1 Anxiety disorder

Anxiety is defined by overwhelming concern and fear, particularly when faced with significant issues or when making important decisions [18]. Anxiety disorder symptoms begin to impact people's daily lives [19]. They experience extreme fear, anxiety, and nervousness [18]. In addition, they have symptoms like a racing heart, difficulty breathing, shivering, and a desire to vomit when confronted with unfavorable circumstances [18]. Anxiety disorders do not only happen to people in certain situations or at certain ages, so anyone can have them [18]. People with GAD have had extreme worry or tension over a variety of topics for at least six months, including personal safety, jobless, social fear, and daily life occurrences [20]. Additionally, GAD sufferers avoid uncertain situations or seek assurance and worry needlessly about adverse outcomes [19]. Extreme fear, an irregular heartbeat, trembling, sweating, shortness of breath, and a sense of losing control are all symptoms of a panic attack [18,20]. The subsequent phobia-related disorder is an extreme phobia of particular things or circumstances [18]. In addition to attempting to escape the event, people with phobia-related conditions have unjustified concerns about a feared object or circumstance.

People who suffer from social anxiety disorder are extremely afraid of what other people will think of their anxious attitudes or behaviors, which makes them feel ashamed [18]. The use of public transportation, being in open places, being in enclosed spaces, waiting in line or being in crowds, and being alone outside the home are among the scenarios in which people with agoraphobia feel utterly terrified [18,20]. People who suffer from separation anxiety disorder are terrified of losing their loved ones, particularly their children [18]. Separation problems could exist in adults as well [20]. They may experience hallucinations of parting from loved ones if separation is occurring or is about to occur [18]. They might endure nightmares as a result of being separated [20]. In addition, they might also display outward signs of anguish that develop over the course of childhood. The signs and symptoms can last into adulthood [20].

5.2.2 Depression disorder

Depression disorders are characterized by sadness, lack of interest or zeal, feelings of guilt or poor self-worth, disturbed sleep or food intake, exhaustion, and reduced focus [20]. Depression, also referred to as clinical depression, is a serious mental health condition that produces severe symptoms that limit one's capacity for feeling, thinking, and carrying out daily activities [21]. In addition, depression can hurt both the depressed individual and others around them. Due of the possibility of suicide, it may be a dangerous medical condition [22]. Constant sadness, a sense of "emptiness," a sense of hopelessness, a loss of interest in activities and hobbies, fatigue, and other symptoms are all indications of depression [21]. There are various forms of depressive disorders, including postpartum depression (PD), psychotic depression, and persistent depressive disorder [22]. A depressed state for two years is referred to as dysthymia or persistent depressive illness [22]. The symptoms of persistent depressive disorder must remain for two years for a person to be diagnosed, even if they experience major depressive episodes and times of less severe symptoms [22].

They experience extreme sadness, anxiety, and exhaustion. Having difficulties handling daily care chores is one of the effects PD has on a mother and her babies [22]. Psychotic symptoms frequently involve a depressive "theme," such as remorse, misery, or illness-related delusions [22]. Typically, the winter months, when little natural sunshine is available, are when seasonal affective disorder first manifests [22]. It is projected that winter depression will revert each year to seasonal affective disorder. Extremely depressed periods that resemble severe depression are standard in people with bipolar illness. Bipolar disorder frequently has euphoric or agitated high moods known as "mania" or a less extreme variation known as "hypomania" [22].

5.2.3 Factors affecting mental health

In general, biological factors, social surroundings, and socioeconomic environments are the root causes of mental health issues [23]. A biological element in mental health issues refers to a faulty nerve cell that links different parts of the brain [23]. Genetics, illnesses that have been linked to brain damage, brain malformations, prenatal injury, and other factors are among the biological causes of mental health problems. For instance, a number of psychiatric problems, such as schizophrenia, have been linked to low birth weight [20]. The phrase "social environment" refers to a person's interactions with their environment and culture. It concerns interpersonal connections with family, close associates, coworkers, and the neighborhood [24]. Socioeconomic status is a reflection of a person's financial situation. The primary causes of mental health issues are financial issues. People in low financial situations frequently exhibit tension and anxiety [24]. Below are listed some of the factors and issues contributing to mental health problems:

- fearful feeling,
- lack of social support,
- being far from home,
- financial problems,
- family problems,
- an unsupportive learning environment,
- childhood adversities.

5.3 Exploratory data analysis and preprocessing

Preprocessing and exploratory data analysis (EDA) comes after data collection. EDA is a strategy that aids in improving data comprehension. This can be accomplished using a variety of visualization techniques, including histograms, scatter plots, and box plots. For instance, in [25], a heatmap was employed for evaluating the differences in activity between participants with and without depression. An EDA can be helpful in locating missing readings brought on by faulty sensors and in detecting outliers. Filtering techniques can be used to lower noise and get rid of outliers. Scaling, quantization, binarization, and other modifications are examples. Dimensionality reduction is another popular preprocessing method that is used to visualize high-dimensional data and minimize the variables to improve the computational efficiency of the model training process. Principal component analysis (PCA) and linear discriminant analysis (LDA) are two methods used frequently to reduce the number of dimensions [26,27]. Fig. 5.1 shows the basic framework of ECG signal analysis and classification.

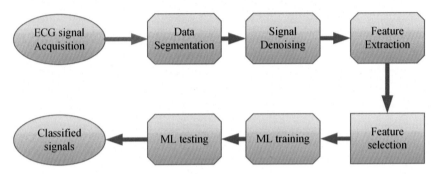

Figure 5.1 The basic framework of ECG signal analysis.

Many machine learning techniques require clear data representations rather than the raw outputs from sensors. One of the most crucial phases in predicting mental states is feature extraction, which is the process of creating a subset of features used in the model. The following features are frequently retrieved for the purpose of identifying mental states, according to the literature [25,28,29]: standard deviation, min, max, skewness, and power spectrum density. This method makes it possible for each sample of a mental state to be represented by a feature vector that matches it. Only a few features in the feature vector will always be important or add much to the forecast. Some feature selection techniques [30] can be used to identify the significance of each feature, hence lowering the data's dimensionality. Others are employed during model training, while some feature selection strategies are utilized prior to model training. (See Table 5.1.)

5.4 Feature extraction

Many physicians have used ECG to diagnose acute or chronic stress. The ECG accurately records the direction and intensity of the electrical current brought about by the atria's and ventricles' depolarization and repolarization. The P-QRS-T waves make up one cardiac cycle in an ECG signal.

Table 5.1 Interrelation among stress, anxiety, and depression [30].

S.No.	SAD (stress, anxiety, depression)	Symptoms
1	Stress [31]	Mild headache, distress, fatigue, worrisome
2	Anxiety [32]	Restlessness, moderate distress, agitation, sweating
3	Depression [33]	Sleeplessness, loss of interest, overthinking, high heart rate
4	All [34]	Health issue, blood pressure, tiredness, EEC signal fluctuation

Fundamental features like amplitudes and intervals are provided by the ECG feature extraction system for use in subsequent automatic analysis. Several methods to identify these features have recently been proposed. The features have been extracted using the following strategy [35]: The highest point in the ECG signal is the R peak. The other features of the signal – Q, S, and T – can be calculated from this point. We locate the highest points, or R peaks, after receiving the ECG signal as a.csv file as input. Then, features Q and S can be found by taking a particular window and finding its lowest points. A window of points prior to the R peak is considered for the Q feature, while a window of points following the R peak is used for the S feature. The T feature is the highest point in the signal after the S feature. The features of an ECG signal can be extracted using a variety of algorithms in a variety of ways. Eigenanalysis is the name of the mathematical approach that is utilized in PCA. One way to extract nonlinear features is through kernel PCA.

5.5 Machine learning-based model for prediction and classification of ECG signals

Data are generated increasingly quickly as a result of the development of information and communication technology. The ability to analyze such vast amounts of data and draw knowledge from them has been made feasible by the rise in computing capacity of machines in recent years. Training a machine learning model involves identifying the parameter values to forecast computational costs and optimize the accuracy rate, model size, final model comprehensibility, etc. Data must be presented in established numerical representations (for example, feature vectors) while training a machine learning model, which is performed during the preprocessing stage. Fig. 5.2

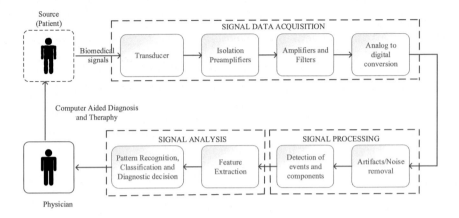

Figure 5.2 Steps to analyze any biomedical signals.

presents the basic ECG signal analysis from data collection and acquisition to signal processing and analysis. Signal processing is the process of altering or modifying data to make sense of elements hidden from direct view. Signal processing allows engineers and scientists to analyze, enhance, and repair signals such as audio streams, images, and video data. System characterization typically involves signal analysis. The use of linear algorithms is the most straightforward strategy for system identification. However, these linear methods may not always produce helpful results depending on the level of nonlinearity of the system at hand. Some applications for machine learning use data and neural networks in a way similar to how a real brain works. Machine learning is also known as predictive analytics for solving business problems.

5.5.1 Different machine learning models

Different types of machine learning algorithms are given below.

5.5.1.1 Supervised learning

In supervised learning, the algorithms are provided a set of training examples (also known as instances). Two input values and one output value make up each instance. The input is frequently shown as an array of numbers that correspond to the instance. The feature vector is the common moniker for this array. The value we wish to forecast, such as a category or a certain number, is the output. The class is frequently used to refer to the outcome in classification-related problems. In the training phase, the algorithms' aim is to figure out how to convert input data into output values. When given fresh, unexpected input values following training, the system will be able to anticipate unknown output values. By "hiding" the output values of the instances and allowing the algorithm to anticipate them, its effectiveness can be evaluated. The outcomes can then be compared with the actual output, often known as the ground truth. Regression is the term used when the output prediction is quantitative and the predictor variable is categorical.

However, the difficulty in predicting property values arises from the fact that the price is a numerical measurement. Regression happens when features like dimensions, position, the number of rooms, etc., are used to calculate the price. A classification model, usually referred to as a classifier, can be used as the results are categorical (stressed and nonstressed). The classifier is the most often used machine learning model for determining mental states. The reason for this is that by allowing the output variable to accept one of the desired mental states, it is possible to formulate the problem of determining whether a particular mental state is present or absent as a classification problem. Some classifiers produce numerical values, such as probabilities, as their output rather than the final category or class. These parameters allow the final category to be decided after that (for example, by choosing the category with the highest probability). There are different classifiers trained through supervised learning which have been used to identify mental states:

- logistic regression,
- random forest,
- naive Bayes,
- decision tree,
- hidden Markov models,
- SVM,
- KNN.

SVM can handle regression and classification on both linear and nonlinear data. Because they are straightforward and easy to understand, decision trees are often used. J48 trees, a form of decision tree, were employed in [36] to classify stress levels, with accuracy rates of 78%. In MHMS, naive Bayes models have also been applied. In [37], they were used to predict bipolar episodes, and in [28] they were used to identify stress.

5.5.1.2 Unsupervised learning

The output variable y is unknown during training; as a result, the aim is to identify the many categories that might spontaneously develop from similarities in the input data, such as identifying user groups who are similar. Unsupervised learning techniques like clustering algorithms are frequently employed to identify groupings or hierarchies in data. Clustering can occasionally be used to prepare data for supervised learning. Clustering methods [38] include:

- K-means,
- hierarchical clustering,
- DBSCAN,
- K-medoids.

For a comprehensive overview of supervised and unsupervised learning, please consult [38,39]. K-means clustering is used to obtain the user groups with similar features and build improved stress detection models for MHMS [28,40].

5.5.1.3 Semisupervised learning

When there are many training data but only a small percentage of samples have the output variable (label) known, this situation is referred to as semisupervised learning. Models are trained using both labeled and unlabeled examples in semisupervised techniques [41]. For the purpose of identifying mental states, semisupervised learning is crucial because it may be difficult to classify the data using the right ground truth type (class). For instance, surveys are typically used to identify daily mood states, but occasionally individuals forget to respond, leading to multiple days going untagged. To tag the data for bipolar disorder, it must be determined whether the patient is now sad, manic, etc. The amount of tagged data that are available will be constrained by problems of this kind. Although semisupervised learning has not been extensively studied for the detection of mental states, certain works have employed it in the past, such as [29] for the prediction of bipolar episodes and [42] for the detection of stress.

5.5.1.4 Transfer learning

Transfer learning is the process of acquiring a new skill in a different domain with little to no labeled training data by using knowledge from related or dissimilar domains with labeled data [43]. For instance, knowledge gained from studying how to recognize horses from photographs may be applied to recognize cows. A more specific illustration is a system that was trained to recognize several image categories based on very fundamental levels, such

Table 5.2 Comparison of different machine learning techniques in predicting depression and anxiety.

Ref.	Methods	Variables	Results
[45]	Random forest (RF), NB, SVM, KNN	Perceived Stress Scale (PSS) questionnaire	RF: 83.33%, 71.42%, 85.71%, and 55.55% respectively
[46]	DT, SVM,	Outgoing activity, toileting activity, sleeping activity, disease, mental status (GDS)	Normal: 95.1%, 75.6%, Mild depression: 94.3%, 75.6%, Severe depression: 99.4%, 99.5%
[47]	KNN, SVM	Depression level, sex, grade, major, technical, novel, amusing, psychological	KNN: 76.6%, SVM: 82%
[48]	Random forest, SVM	Age, MMSE score, depression (GDS), MoCA test	RF: 95.45%, SVM: 92.42%
[49]	XGBoost	Sociodemographic information, earthquake-related experience, sleep, mood	Classical accuracy of 74.476%
[50]	SVM, KNN,	Standard deviation of heart rate, energy of ECG	Accuracy of SVM 82.5%, KNN 87.6%

as cats, dogs, cars, etc., and then develop the ability to identify whether or not an image has a particular condition [44]. (See Table 5.2.)

This strategy has the major benefit of helping to solve the issue of not having enough labeled data. It appears to perform fairly well, especially for use cases involving images [43]. The findings are poor if the domains are too dissimilar, which is a drawback. Additionally, it has not been thoroughly investigated in contexts other than pictures and movies. Transfer learning with decision trees incorporated was employed in [42] for stress detection where there is a paucity of labeled data. Deep transfer learning and a convolutional network were used in [51] to perform early Alzheimer's disease diagnosis utilizing input of images of brains.

5.5.1.5 *Reinforcement learning*

Through interactions with its surroundings, an agent learns by doing via reinforcement learning [52]. The agent's goal is to get the most rewards possible and choose the best course of action given the situation. In contrast to supervised learning, where the inputs and goal outputs are supplied to the learning algorithm, this involves an agent exploring potential actions and choosing the one that will maximize the reward. An agent is rewarded or penalized appropriately depending on whether it engages in proper or destructive activity. It attempts to learn over time what behavior yields the best reward. One benefit of this algorithm is that a human expert knowledgeable in the problem domain is optional. Applications of reinforcement learning have been used in healthcare before. For example, the authors of [53] used reinforcement learning to reduce the number and length of seizures in people with epilepsy. Various algorithms are frequently mixed in practice to create the final prediction models. Additionally, there are two types of training methods: user-dependent and user-independent (general) models. The first type can only be trained using the information provided by the user. These are prepared using data from users other than the target user, who is the person who is supposed to use the system. User-dependent models can accurately capture each user's unique behavior and frequently produce better results. Still, they demand a large amount of training data for that single user.

5.6 Conclusion and future scope

In the present chapter, existing literature is reviewed on stress detection to examine the relationship between psychological activity and numerous bodily signs, including galvanic skin resistance, skin conductance, skin temperature, heart rate, and facial and vocal cues. The dataset from the aforementioned questionnaire and its related relevance have also been investigated. The chapter also provides a brief overview of stress-related variables and symptoms. The chapter highlights the value of machine learning-based AI algorithms in identifying stress. We analyzed the most recent research on monitoring mental status using sensors to collect behavior data and machine learning to understand it. According to the reviewed literature, combining machine learning techniques with multimodal sensing technologies will greatly improve technology-assisted mental healthcare development. Future research about the use of optimization tools to discover a workable design and solution would be fascinating. Our study is limited by the fact that we have not yet used any optimization techniques to improve the model parameters and depression detection, but we think that once we do, the performance of the suggested solution would improve even more.

Young adults between the ages of 15 and 24 are experiencing a sharp rise in mental health difficulties in recent years, such as panic attacks and depression. The rise in patients with mental health conditions necessitates the development of efficient diagnosis techniques that enable rapid and precise evaluation of the patient's mental state. The main goal of the monitoring system is to periodically check the individual's ECG and galvanic skin resistance data in order to generate a report on their mental health status. Although the existing approach can use a person's usual emotional state to infer the existence of a mental health problem like anxiety, there are many more mental health issues, such as depression, bipolar illness, and autism spectrum disorder. The system may be enhanced to detect and keep track of the potential for those mental health problems as well.

References

[1] S. Bhadra, C.J. Kumar, An insight into diagnosis of depression using machine learning techniques: a systematic review, Current Medical Research and Opinion 38 (5) (2022) 749–771.

[2] M. de Bardeci, C.T. Ip, S. Olbrich, Deep learning applied to electroencephalogram data in mental disorders: a systematic review, Biological Psychology 162 (2021) 108117.

[3] A. Alqudah, A. Al-Smadi, M. Oqal, E.Y. Qnais, M. Wedyan, M. Abu Gneam, R. Alnajjar, M. Alajarmeh, E. Yousef, O. Gammoh, About anxiety levels and anti-anxiety drugs among quarantined undergraduate Jordanian students during COVID-19 pandemic, International Journal of Clinical Practice 75 (7) (2021) e14249.

[4] S. Amendola, A. von Wyl, T. Volken, A. Zysset, M. Huber, J. Dratva, A longitudinal study on generalized anxiety among university students during the first wave of the COVID-19 pandemic in Switzerland, Frontiers in Psychology 12 (2021) 643171.

[5] L.L. Fahlberg, L.A. Fahlberg, Exploring spirituality and consciousness with an expanded science: beyond the ego with empiricism, phenomenology, and contemplation, American Journal of Health Promotion 5 (4) (1991) 273–281.

[6] S.L. Prescott, A.C. Logan, D.L. Katz, Preventive medicine for person, place, and planet: revisiting the concept of high-level wellness in the planetary health paradigm, International Journal of Environmental Research and Public Health 16 (2) (2019) 238.

[7] Z. Bagheri, P. Noorshargh, Z. Shahsavar, P. Jafari, Assessing the measurement invariance of the 10-item Centre for Epidemiological Studies Depression Scale and Beck Anxiety Inventory questionnaires across people living with HIV/AIDS and healthy people, BMC Psychology 9 (2021) 1–11.

[8] R.C. Kessler, H.M. van Loo, K.J. Wardenaar, R.M. Bossarte, L. Brenner, D. Ebert, P. De Jonge, A. Nierenberg, A. Rosellini, N. Sampson, et al., Using patient self-reports to study heterogeneity of treatment effects in major depressive disorder, Epidemiology and Psychiatric Sciences 26 (1) (2017) 22–36.

[9] H.-U. Wittchen, F. Jacobi, J. Rehm, A. Gustavsson, M. Svensson, B. Jönsson, J. Olesen, C. Allgulander, J. Alonso, C. Faravelli, et al., The size and burden of mental

disorders and other disorders of the brain in Europe 2010, European Neuropsychopharmacology 21 (9) (2011) 655–679.

[10] S.M. Alarcao, M.J. Fonseca, Emotions recognition using EEG signals: a survey, IEEE Transactions on Affective Computing 10 (3) (2017) 374–393.

[11] J. Taelman, S. Vandeput, A. Spaepen, S. Van Huffel, Influence of mental stress on heart rate and heart rate variability, in: 4th European Conference of the International Federation for Medical and Biological Engineering: ECIFMBE, 23–27 November 2008, Antwerp, Belgium, Springer, 2009, pp. 1366–1369.

[12] E. Ertin, N. Stohs, S. Kumar, A. Raij, M. Al'Absi, S. Shah, AutoSense: unobtrusively wearable sensor suite for inferring the onset, causality, and consequences of stress in the field, in: Proceedings of the 9th ACM Conference on Embedded Networked Sensor Systems, 2011, pp. 274–287.

[13] M. Wu, Y. Lu, W. Yang, S.Y. Wong, A study on arrhythmia via ECG signal classification using the convolutional neural network, Frontiers in Computational Neuroscience 14 (2021) 564015.

[14] S. Anjume, K. Amandeep, A. Aijaz, F. Kulsum, Performance analysis of machine learning techniques to predict mental health disorders in children, International Journal of Innovative Research in Computer and Communication Engineering 5 (5) (2017).

[15] N.S.M. Shafiee, S. Mutalib, Prediction of mental health problems among higher education student using machine learning, International Journal of Education and Management Engineering (IJEME) 10 (6) (2020) 1–9.

[16] S.M. Awaluddin, et al., National Health and Morbidity Survey (NHMS) 2017: Key findings from the adolescent health and nutrition surveys, National Institute of Health, Ministry of Health Malaysia, 2018.

[17] D.C. Gonçalves, G.J. Byrne, Interventions for generalized anxiety disorder in older adults: systematic review and meta-analysis, Journal of Anxiety Disorders 26 (1) (2012) 1–11.

[18] G. Andrews, C. Bell, P. Boyce, C. Gale, L. Lampe, O. Marwat, R. Rapee, G. Wilkins, Royal Australian and New Zealand college of psychiatrists clinical practice guidelines for the treatment of panic disorder, social anxiety disorder and generalised anxiety disorder, Australian & New Zealand Journal of Psychiatry 52 (12) (2018) 1109–1172.

[19] S.M.A. Shah, D. Mohammad, M.F.H. Qureshi, M.Z. Abbas, S. Aleem, Prevalence, psychological responses and associated correlates of depression, anxiety and stress in a global population, during the coronavirus disease (COVID-19) pandemic, Community Mental Health Journal 57 (2021) 101–110.

[20] World Health Organization, Depression and other common mental disorders: global health estimates (No. WHO/MSD/MER/2017.2), World Health Organization, 2017.

[21] B.A. Teachman, D. McKay, D.M. Barch, M.J. Prinstein, S.D. Hollon, D.L. Chambless, How psychosocial research can help the national institute of mental health achieve its grand challenge to reduce the burden of mental illnesses and psychological disorders, The American Psychologist 74 (4) (2019) 415.

[22] K. Bustamante, Developing information on mental health and counseling services, https://digitalcommons.csumb.edu/caps_thes_all/527, 2019.

[23] S. O'Neill, M. McLafferty, E. Ennis, C. Lapsley, T. Bjourson, C. Armour, S. Murphy, B. Bunting, E. Murray, Socio-demographic, mental health and childhood adversity risk factors for self-harm and suicidal behaviour in College students in Northern Ireland, Journal of Affective Disorders 239 (2018) 58–65.

[24] I.L.D. Moutinho, A.L.G. Lucchetti, O. da Silva Ezequiel, G. Lucchetti, Mental health and quality of life of Brazilian medical students: incidence, prevalence, and associated factors within two years of follow-up, Psychiatry Research 274 (2019) 306–312.

[25] E. Garcia-Ceja, M. Riegler, P. Jakobsen, J. Tørresen, T. Nordgreen, K.J. Oedegaard, O.B. Fasmer, Depresjon: a motor activity database of depression episodes in unipolar

and bipolar patients, in: Proceedings of the 9th ACM Multimedia Systems Conference, 2018, pp. 472–477.

[26] L. Rebollo-Neira, D. Černá, Wavelet based dictionaries for dimensionality reduction of ECG signals, Biomedical Signal Processing and Control 54 (2019) 101593.

[27] J.C. Gower, Some distance properties of latent root and vector methods used in multivariate analysis, Biometrika 53 (3–4) (1966) 325–338.

[28] E. Garcia-Ceja, V. Osmani, O. Mayora, Automatic stress detection in working environments from smartphones' accelerometer data: a first step, IEEE Journal of Biomedical and Health Informatics 20 (4) (2015) 1053–1060.

[29] A. Maxhuni, A. Muñoz-Meléndez, V. Osmani, H. Perez, O. Mayora, E.F. Morales, Classification of bipolar disorder episodes based on analysis of voice and motor activity of patients, Pervasive and Mobile Computing 31 (2016) 50–66.

[30] G. Chandrashekar, F. Sahin, A survey on feature selection methods, Computers & Electrical Engineering 40 (1) (2014) 16–28.

[31] Y. Liu, X. Lv, Z. Tang, The impact of mortality salience on quantified self behavior during the COVID-19 pandemic, Personality and Individual Differences 180 (2021) 110972.

[32] W. Hart, C. Kinrade, M. Xia, J.T. Lambert, The positive-passion hypothesis: grandiose but not vulnerable narcissism relates to high-approach positive affect following provocation, Personality and Individual Differences 180 (2021) 110983.

[33] K. Yin, P. Lee, O.J. Sheldon, C. Li, J. Zhao, Personality profiles based on the FFM: a systematic review with a person-centered approach, Personality and Individual Differences 180 (2021) 110996.

[34] A.K. Winiker, K.E. Tobin, M. Davey-Rothwell, C. Latkin, An examination of grit in black men who have sex with men and associations with health and social outcomes, Journal of Community Psychology 47 (5) (2019) 1095–1104.

[35] P.K. Sahoo, H.K. Thakkar, W.-Y. Lin, P.-C. Chang, M.-Y. Lee, On the design of an efficient cardiac health monitoring system through combined analysis of ECG and SCG signals, Sensors 18 (2) (2018) 379.

[36] D. Carneiro, J.C. Castillo, P. Novais, A. Fernández-Caballero, J. Neves, Multimodal behavioral analysis for non-invasive stress detection, Expert Systems with Applications 39 (18) (2012) 13376–13389.

[37] A. Grünerbl, A. Muaremi, V. Osmani, G. Bahle, S. Oehler, G. Tröster, O. Mayora, C. Haring, P. Lukowicz, Smartphone-based recognition of states and state changes in bipolar disorder patients, IEEE Journal of Biomedical and Health Informatics 19 (1) (2014) 140–148.

[38] I.H. Witten, E. Frank, M.A. Hall, C.J. Pal, M. Data, Practical machine learning tools and techniques, in: Data Mining (Vol. 2, No. 4), Elsevier, Amsterdam, The Netherlands, June 2005, pp. 403–413.

[39] S. Aziz, S. Ahmed, M.-S. Alouini, ECG-based machine-learning algorithms for heartbeat classification, Scientific Reports 11 (1) (2021) 18738.

[40] Q. Xu, T.L. Nwe, C. Guan, Cluster-based analysis for personalized stress evaluation using physiological signals, IEEE Journal of Biomedical and Health Informatics 19 (1) (2014) 275–281.

[41] X. Zhai, Z. Zhou, C. Tin, Semi-supervised learning for ECG classification without patient-specific labeled data, Expert Systems with Applications 158 (2020) 113411.

[42] A. Maxhuni, P. Hernandez-Leal, L.E. Sucar, V. Osmani, E.F. Morales, O. Mayora, Stress modelling and prediction in presence of scarce data, Journal of Biomedical Informatics 63 (2016) 344–356.

[43] S.J. Pan, Q. Yang, A survey on transfer learning, IEEE Transactions on Knowledge and Data Engineering 22 (10) (2010) 1345–1359.

[44] K. Pogorelov, M. Riegler, S.L. Eskeland, T. de Lange, D. Johansen, C. Griwodz, P.T. Schmidt, P. Halvorsen, Efficient disease detection in gastrointestinal videos–global features versus neural networks, Multimedia Tools and Applications 76 (2017) 22493–22525.

[45] A.A. Sabourin, J.C. Prater, N.A. Mason, Assessment of mental health in doctor of pharmacy students, Currents in Pharmacy Teaching and Learning 11 (3) (2019) 243–250.

[46] Y. Hou, J. Xu, Y. Huang, X. Ma, A big data application to predict depression in the university based on the reading habits, in: 2016 3rd International Conference on Systems and Informatics (ICSAI), IEEE, 2016, pp. 1085–1089.

[47] I.-M. Spyrou, C. Frantzidis, C. Bratsas, I. Antoniou, P.D. Bamidis, Geriatric depression symptoms coexisting with cognitive decline: a comparison of classification methodologies, Biomedical Signal Processing and Control 25 (2016) 118–129.

[48] F. Ge, Y. Li, M. Yuan, J. Zhang, W. Zhang, Identifying predictors of probable post-traumatic stress disorder in children and adolescents with earthquake exposure: a longitudinal study using a machine learning approach, Journal of Affective Disorders 264 (2020) 483–493.

[49] A. Hasanbasic, M. Spahic, D. Bosnjic, V. Mesic, O. Jahic, et al., Recognition of stress levels among students with wearable sensors, in: 2019 18th International Symposium INFOTEH-JAHORINA (INFOTEH), IEEE, 2019, pp. 1–4.

[50] H.S. AlSagri, M. Ykhlef, Machine learning-based approach for depression detection in Twitter using content and activity features, IEICE Transactions on Information and Systems 103 (8) (2020) 1825–1832.

[51] S. Basu, K. Wagstyl, A. Zandifar, L. Collins, A. Romero, D. Precup, Early prediction of Alzheimer's disease progression using variational autoencoders, in: Medical Image Computing and Computer Assisted Intervention–MICCAI 2019: 22nd International Conference, Proceedings, Part IV 22, October 13–17, 2019, Shenzhen, China, Springer, Shenzhen, China, 2019, pp. 205–213.

[52] A. Coronato, M. Naeem, G. De Pietro, G. Paragliola, Reinforcement learning for intelligent healthcare applications: a survey, Artificial Intelligence in Medicine 109 (2020) 101964.

[53] A. Insani, W. Jatmiko, A.T. Sugiarto, G. Jati, S.A. Wibowo, Investigation reinforcement learning method for R-wave detection on electrocardiogram signal, in: 2019 International Seminar on Research of Information Technology and Intelligent Systems (ISRITI), IEEE, 2019, pp. 420–423.

CHAPTER 6

A robust peak detection algorithm for localization and classification of heart sounds in PCG signals

Shrey Agarwal[a], Yashaswi Upmon[b], Muhammad Zubair[c], and Umesh Kumar Naik Mudavath[d]

[a]Department of Computer Science, University of Southern California, Los Angeles, CA, United States
[b]Department of Computer Science Engineering, Kalinga Institute of Industrial Technology, Bhubaneshwar, Orissa, India
[c]Department of Biomedical Engineering, The Pennsylvania State University, State College, PA, United States
[d]Department of Electronics and Electrical Engineering, Indian Institute of Technology Guwahati, North Guwahati, Assam, India

Contents

6.1 Introduction

As per the Centers for Disease Control and Prevention (CDC), heart disease is the major cause of mortality, accounting for 659,041 deaths in the US per year [1]. To detect heart diseases with the help of automated algorithms, firstly, we need to understand the four fundamental heart sounds S1, S2, S3, and S4. The first heart sound, S1, is produced by the closing of the mitral and tricuspid valves. The second heart sound, S2, results from

Signal Processing Driven Machine Learning Techniques for Cardiovascular Data Processing
https://doi.org/10.1016/B978-0-44-314141-6.00011-6

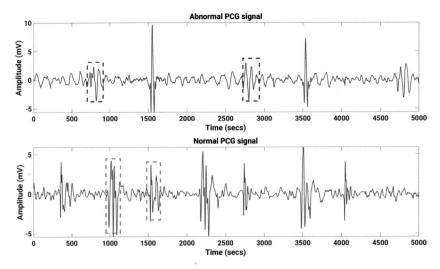

Figure 6.1 Positions of S3 and S4 (red) and S1 and S2 (green) heart sounds in an abnormal/normal PCG signal.

the closure of the aortic and pulmonic valves. The third heart sound, S3, also termed ventricular gallop, is heard just after S2 when the mitral valve opens to allow passive filling of the left ventricle (LV). Lastly, the fourth heart sound, S4, also known as the atrial gallop, comes just before S1 when the atria contract to force blood into the LV [2].

The quickest, simplest, and most economical initial line of screening is cardiac auscultation. It is performed for patients with heart problems, including valve disease, arrhythmia, heart failure, pulmonary hypertension, etc. Contrary to this, heart sounds are tough to recognize and examine by physicians because the important events are closely positioned in time, the sounds are feeble, and their frequency components are at the bottom end of the audible frequency span [3]. Large-scale tutoring is needed to grasp cardiac auscultation adequately. Only 20% of medical trainees can successfully detect cardiac abnormalities using auscultation, according to the findings reported in [4].

Computerized pathological heart sound localization can be considered a crucial step in the detection of cardiovascular disease in the field of biomedical engineering [5]. The process of detecting whether a heart sound is normal or pathological can be broken down into three essential steps: localization, feature extraction, and classification [6]. Firstly, the localization technique is used to locate the elementary heart sounds (i.e., S1 and S2) of the PCG signal, as illustrated in Fig. 6.1. But predicting the elementary heart sounds is altogether a composite task and the sound can be altered by other sounds, such as the presence of S3 and S4 heart sounds and artifacts like murmurs [7]. Previously, Messener et al. [8] extracted the envelope and spectral features for detecting abnormal and normal sequences using a deep recurrent neural network (RNN). In [9], by using ensemble empirical mode decomposition (EEMD), the authors extracted the kurtosis feature to localize the presence of S1 and S2 heart sounds. Springer et al. [10] proposed an algorithm for correct segmentation of the first and second heart sound in the PCG signals using a hidden semi-Markov model (HSMM) and logistic regression. In [11], the Hilbert vibration decomposition system is used to decompose the heart sound into a fixed number of subcomponents while conserving the phase information further to expedite the localization and identification of the signal. However, the performance of the algorithms reported in these works is rather subpar when compared to the latest state-of-the-art fully automated algorithms.

In order to improve the classification accuracy and automate the feature selection process, we proposed a robust algorithm in which we used the labeled peaks of PCG signals extracted using wavelet-based K-means clustering as the features for the various machine learning (ML)/deep learning (DL) classifiers to recognize normal and abnormal heart sounds. Our work presents a novel approach using an unsupervised learning model named wavelet-based K-means clustering, which has been implemented on the PhysioBank ATM Challenge 2016 database for the localization of heart sounds. Being unsupervised, it may be able to recognize the fundamental sounds in real-time noisy PCG signals. Also, it saves time by using the labeled peaks as the feature for the classification models instead of manually finding the features of the peaks. We have attained trailblazing outcomes compared with the start-of-the-art methods for numerous databases present in PhysioBank ATM Challenge 2016. For a better understanding of abnormal and normal heart sounds, normal S1 and S2 and abnormal S3 and S4 sounds are shown in Fig. 6.1.

6.2 Literature review

Much of the prior work was focused on PCG segmentation using DL and ML algorithms. In [12], Fernando et al. proposed a model to detect the heart state from the PCG signals by using RNNs. They have used attention-based learning for the extraction of salient features from noisy and irregular heart signals. Their proposed algorithm was implemented on two databases, namely, "PhysioNet/CinC Challenge 2016 (PCC)" and the "M3dicine database," which is a privately collected dataset consisting of both human and animal heart sounds. For segmentation tasks, numerous feature combinations were evaluated.

Dissanayake et al. [13] exploited the segmentation of heart sound before proceeding with classification by using three different DL architectures. In [14], the proposed procedure uses shapley additive explanations (SHAP) and an occlusion map algorithm, which aid the segmentation to play a vital role in the classification and detection of abnormal and normal heart sounds. Further, the use of MFCC features can easily identify the location of S1 and S2 in PCG signals. The new proposed classifiers in the above work can classify the heart signals with an accuracy of approximately 100%.

Renna et al. [15] employed a deep convolutional neural network (CNN) in concurrence with the underlying hidden Markov models (HMM) and HSMM to deduce emission distribution. The outcome of the work has given an average sensitivity of 93% along with an average true positive rate of 94%.

Gojoreski et al. [16] described an approach for the prediction of chronic heart failure using PCG signals. In this method, classical ML models have been trained from numerous expert-designed features, whereas DL models have been trained from a spatiotemporal and spectral representation of signals. The model has then been implemented on their chronic heart failure (CHF) database along with an open source training dataset available at "PhysioNet Cardiology Challenge." This methodology attains 89.3% accuracy compared to the baseline model. They infer 15 expert features that can be useful for the detection of CHF phases with a claimed accuracy of 93.2%.

Naami et al. [17] described an adaptive network-based fuzzy inference system (ANFIS) for the classification of abnormal signals. The denoising of the PCG signals has been performed using notch and Butterworth filtering, and the denoised signal is then fed to discrete Fourier transform (DFT), and from the third cumulant using high-order spectral (HOS) analysis, five features were extracted. These five features have been used with neural network systems to classify heart signals into abnormal and normal signals. The accuracy of their proposed model ranges from 63% to 89% for the classification task.

Tien-En Chen et al. [18] proposed a DNN-based method for the detection of S1 and S2 heart sounds. The sounds are converted to MFFCs.

The MFFCs features are clustered into two categories, which are then fed into DNN, which further classifies S1 and S2 with an accuracy of 91%. Mishra et al. [19] considered morphological characteristics derived from the PCG signals for recognition of S1 and S2. This method is implemented on publicly available databases, along with which they have implemented this on experimentally collected HSS recordings under the supervision of the physicians. Quantifying the nonlinear and nonstationary nature of PCG signals was inferred from variational mode decomposition (VMD), which is based on quantifying the nonstationary and nonlinear nature of PCG signals. In comparison to previous works, their experiments show a significant improvement in the classification task.

All the prior works discussed above are based on the classification of heart sounds in PCG signals. Most of them are focused on supervised learning for signal localization and classification. Moreover, DL models require an enormous amount of data for training, which is a time-consuming task and increases the overall complexity. As is well known, while analyzing biomedical data, the major difficulty is handling class imbalance. To resolve this constraint, we have proposed a novel algorithm using unsupervised learning that is time-efficient, accurate, and precise in both the localization and classification of heart sounds. Additionally, most of the other works have used feature extraction techniques, which makes the process more computationally expensive and laborious. Therefore, in our work, we have not statistically calculated the features and have instead used an automated process for finding the peaks in the signal, which further acts as features for the classification models. To achieve the best performance, we have extracted the high-frequency or detailed components (DCs) using DWT, which contains most of the useful information related to heart sounds. This also helps us to denoise the PCG signal before giving the wavelets to the proposed unsupervised model. When compared to the above state-of-the-art works, simulation results show that our suggested approach outperforms the previously reported models.

6.3 Proposed methodology

Firstly, we collect the PCG data from PhysioBank ATM. Secondly, we divide the PCG signal into various high-frequency components or detailed components using DWT. Since the heart sounds are found in the high-frequency components, this move to use DWT helps us to identify legit peaks using the different wavelets. Then, we compute the absolute value

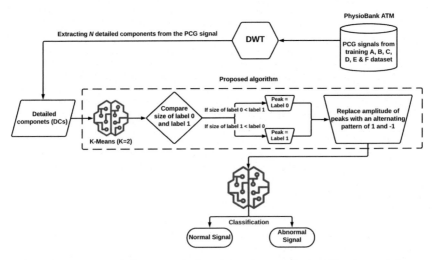

Figure 6.2 Proposed framework for detection and classification of heart sounds from PCG signals.

of the wavelet amplitude, which is applied as an input to the K-means clustering algorithm. This unsupervised algorithm clusters the amplitude of the PCG signals into two clusters and labels them as 0 and 1, which indicate abnormal and normal peaks of PCG signals, respectively. Further, to identify the cluster having peaks, we compare the size of the clusters. Whichever cluster is smaller in size, we consider it to contain the peaks of the signals. Then, we label those peaks using a pattern of -1 and 1, and the rest of the signal is labeled as 0. Next, we balance this dataset using the synthetic minority oversampling technique (SMOTE) to overcome the problem of class imbalance. Later we use this pattern to predict abnormal and normal heart sounds using various ML/DL algorithms. The system architecture of the proposed methodology is depicted in Fig. 6.2, and a detailed explanation is given in the following subsections.

6.3.1 Discrete wavelet transform

Wavelet transform (WT) is a suitable evaluation tool for nonstationary signals like PCG after the complete signal is represented in matrix form. In this work, WT is employed to decompose the PCG signal into diverse frequency band signals. After processing, the signal is reconstructed by excluding the noncritical part of the signal. DWT is an efficient way to analyze the PCG signal, which can also reduce the time complexity. Fig. 6.3 illustrates all eight levels of frequency components retrieved em-

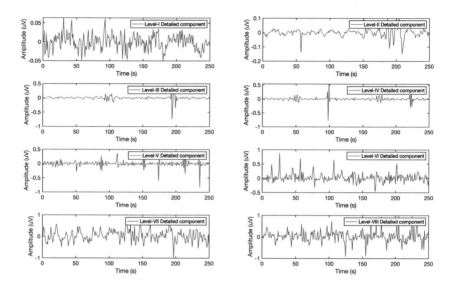

Figure 6.3 Output of detailed components extracted using 8-level DWT.

ploying DWT. The signal is decomposed into finer details by employing two functions, as illustrated below, in DWT:

$$D(t) = \sum_{l \in Z} 2^{\frac{j}{2}} a_j(k)\varphi\left(2^j t - k\right) + \sum_{j=0}^{j-1} \sum_{k=0}^{\infty} 2^{\frac{j}{2}} d_j(k)\psi\left(2^j t - k\right), \quad (6.1)$$

$$a_j(k) = \int_{-\infty}^{\infty} D(t)\varphi\left(2^j t - k\right) dt, \quad (6.2)$$

$$d_j(k) = \int_{-\infty}^{\infty} D(t)\psi\left(2^j t - k\right) dt, \quad (6.3)$$

where $\varphi(t)$ is a basic scaling function and $\psi(t)$ is the mother wavelet. In (6.2) and (6.3), a_j and d_j are approximate and detailed coefficients of the decomposed signal, respectively. Initially, the PCG signal is split into two different named components: a high-frequency component (HFC) (d_j) and a low-frequency component (LFC) (a_j). By downsampling the PCG signal by a factor of two, we obtain the approximation of LFC and HFC. We repeat this process until the desired decomposition level is reached (i.e., eight in our case).

6.3.2 Peak detection using K-means clustering

For a normal cardiac signal, the cycle consists of four major regions, that is, the first heart sound (S1), the systole pause interval (SPI), the second heart sound (S2), and the diastole pause interval [19]. Pathological signals may contain a third heart sound (S3) and a fourth heart sound (S4) [20]. To identify the four sounds using ML models, we use an unsupervised K-means clustering method. We consider both abnormal and normal datasets for the implementation of the proposed methodology. The representation of one of the normal signals and abnormal signals present in the dataset is depicted in Fig. 6.1. Next, to identify the peaks in PCG signals, we propose a novel algorithm that is based on K-means clustering as described in Algorithm 6.1. In Algorithm 6.1, we first take the absolute value or modulus of the amplitude of the PCG signals. Then, we implement K-means clustering where the value of k is taken as 2. The equation of K-means clustering is as follows [21]:

$$J = \sum_{j=1}^{k} \sum_{i=1}^{n} \|x_i^j - c_j\|^2, \tag{6.4}$$

where J is the objective function, k is the number of clusters, n is the number of cases, x_i is the ith case, and c_j is the centroid for cluster j. The Euclidean distance is calculated by the function $\|x_i^j - c_j\|^2$, whose sum is for all the amplitudes of the PCG signal. The distance is summed up for both centroids. Then we consider the data points for the centroid, which has the minimum distance.

As there are two clusters, the cluster with more data points is suppressed as it does not contain the peaks. This can be inferred by looking at the signal that the data points of peaks in the signal are fewer than the data points in the rest of the signal. To suppress the signal, we substitute all the nonpeak signals with 0 and peaks with a pattern of 1 and −1. The representation of 1 and −1 depends on the representation of the latest peak example; if 1 has already been assigned to the previous peak, then −1 is assigned to the current peak. After running the peak detection algorithm for both abnormal and normal PCG signals, we consider the new values of the peak as mentioned in step 5 of Algorithm 6.1 for the classification of the signals. The detected peaks of abnormal and normal PCG signals are depicted in Fig. 6.4.

Algorithm 6.1: Peak Detection Algorithm using K-means Clustering.

A = Level 7 DC extracted using DWT
X = abs(A)
zero = 0
one = 0
switch = -1
K-means clustering (number of clusters = 2)
for value in K-means output **do**
  if value == 1 **then**
    one = one + 1
  else
    zero = zero + 1

    if one > zero **then**
      non-peak = one
    else
      non-peak = zero
    end if
    for each signal in dataset **do**
      if switch == -1 **then**
        switch = 1
      else
        switch = -1
      end if
      if value of signal == non-peak **then**
        signal = 0
      else
        signal = switch
      end if
    end for

6.4 Experimental results

6.4.1 Dataset description

The proposed methodology is implemented and evaluated on the Challenge 2016 training set A, B, C, D, E, and F, all taken together as one dataset from the PhysioBank database [22]. Data have been collected from

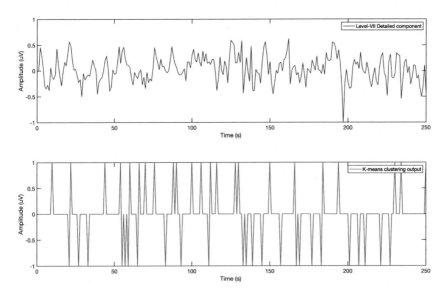

Figure 6.4 Peak detection algorithm applied on the Level-VII Detailed component of the PCG signal.

both healthy subjects and unhealthy patients in clinical or nonclinical settings, and the heart sounds were recorded from different sites on the body. In addition, the normal signals were acquired from healthy patients, and the abnormal recordings were acquired from subjects with established cardiac problems. Both the healthy patients and the unhealthy subjects included adults and children. The important point is that the dataset is imbalanced, meaning that the number of normal signals does not match that of abnormal signals. Since the recordings were made in an uncontrolled environment, most of the recordings are disturbed by various noises.

6.4.2 Peak localization

In this work, we have used DWT to remove the noise from the signal. Then, Algorithm 6.1 was applied to the output of DWT. In this process, the denoised signals are clustered, and the amplitude of the signals is replaced with 0, −1, and 1. Here, 1 and −1 represent the peak values, and the rest of the signal is represented by 0.

6.4.3 Classification results

Based on the features retrieved from the proposed approach, five models were implemented for the classification of abnormal and normal heart

Table 6.1 Evaluation of classification models.

Model	Accuracy	Precision	Recall	F1-Score
KNN	92	0.87	0.88	0.86
GBoost	100	1.0	1.0	1.0
AdaBoost	100	1.0	1.0	1.0
SVM	100	1.0	1.0	1.0
1D CNN	100	0.42	0.50	0.46
LSTM	84	0.42	0.50	0.46

sounds from the PCG signals. The target value is a normal or abnormal signal, where an abnormal signal is considered 1 and a normal signal is considered 0. We used a grid search CV to tune the hyperparameters. The cross-validation value was set to 10. The models employed for classification are K-nearest neighbors, Gradient Boosting, AdaBoost, SVM, 1D CNN, and LSTM.

6.5 Results and discussion

In this work, the classification task has been evaluated on various metrics, i.e., precision, accuracy, F1-score, and recall. The results are depicted in Table 6.1. To further validate and visualize the performance, we have plotted the ROC curve for the SVM classifier because it achieved the highest average accuracy, as depicted in Fig. 6.6. The histogram of performance metrics of classification models is illustrated in Fig. 6.5, and the average accuracy and precision of training sets A, B, C, and F are illustrated in Fig. 6.7. We have also shown the average precision, average recall, average accuracy, and average F1-score in Table 6.2. It is observed that the SVM classifier has demonstrated higher classification performance as compared to other machine learning techniques for classifying normal and abnormal PCG signals. The clustering and classification of abnormal and normal signals are implemented using python 3.8.5.

Table 6.3 presents the accuracy of SVM in classifying PCG signals as normal or abnormal, compared with other state-of-the-art methods. Noman et al. [23] extended the MSAR model to a Switching Linear Dynamic System (SLDS) and developed a new algorithm using the duration-dependent Viterbi algorithm and switching Kalman filter, achieving an accuracy of 91.2% on the same dataset. Humanyu et al. [24] proposed a CNN layer with time-convolutional units, which they evaluated on publicly available multidomain datasets, achieving an accuracy of 82.27%.

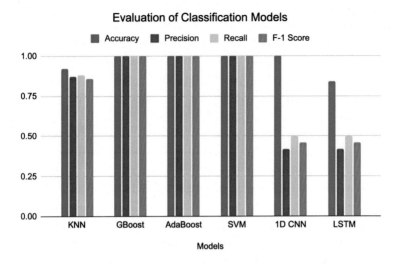

Figure 6.5 Evaluation of classification models.

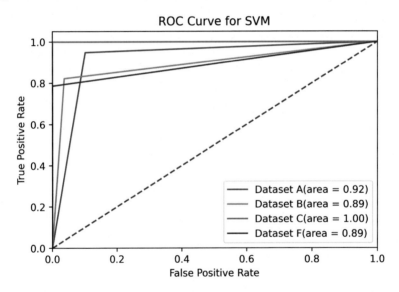

Figure 6.6 The ROC curve for the SVM classifier.

Xiao et al. [25] implemented a DL–based heart sound classification system for cardiovascular disease prediction and obtained an accuracy of 93% on the same dataset. Al-Naami et al. [17] used ANFIS with spectral analysis features to detect abnormal heart sounds and achieved a classification accuracy ranging from 63% to 89%. Shuvo et al. [26] implemented Car-

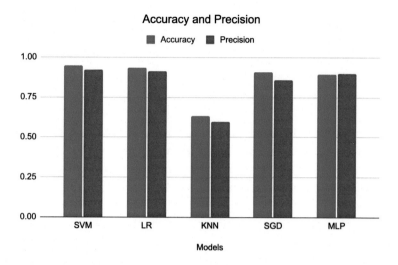

Figure 6.7 Average accuracy and precision of training sets A, B, C, and F.

Table 6.2 Average of metrics of training sets A, B, C, and F.

Model	Accuracy	Precision	Recall	F1-Score
SVM	0.94	0.92	0.91	0.9
LR	0.93	0.91	0.91	0.92
KNN	0.63	0.6	0.59	0.52
SGD	0.90	0.86	0.85	0.85
MLP	0.89	0.9	0.89	0.89

dioXNet, a new CRNN structure for automatic recognition of five target values of cardiac auscultation, obtaining an accuracy of 86% on the same dataset.

Boulares et al. [27] fine-tuned pretrained CNN models, achieving an accuracy of 89%. Chen et al. [28] proposed a system that combines CNN and modified frequency slice wavelet transform (MFSWT) for heart sound classification, achieving an accuracy of 93.91% using PhysioNet Computing in Cardiology Challenge 2016. Li et al. [29] combined standard feature engineering methods with DL algorithms, achieving an accuracy of 86.8% using stratified 5-fold cross-validation on the PhysioNet/CinC Challenge 2016 dataset.

Furthermore, Ghosh et al. [30] developed algorithms using time-frequency-domain (TFD) DNN and SVM to detect heart sound activity from PCG signals, obtaining accuracies of 95.25% and 95.43% on the PhysioNet challenge 2016 (PNC) database, respectively. They also proposed an algorithm to detect heart valve disorders (HVDs) from PCG signals using wavelet synchrosqueezing transform (WST) combined with a random forest classifier, achieving an average accuracy of 95.12% [31]. Similarly, Karhade et al. [32] proposed a TFD-based DL framework for HVD detection using PCG signals, obtaining an accuracy of 85.16% on the PNC database.

From the above discussed work, it can be concluded that the proposed algorithm can achieve superior accuracy and works without taking any features of the fundamental heart sounds for classification. Moreover, the proposed approach uses an unsupervised ML method for the localization of heart sounds, which does not need to know the location of signals beforehand nor requires any preprocessing of the data. The proposed work presents several advantages and disadvantages. The main advantage of the proposed approach is its ability to develop a fully automated approach for localizing heart sounds without requiring extensive preprocessing of the data. This approach uses wavelet-based K-means clustering to distinguish between primitive heart sounds, such as S1, S2, S3, and S4, and inconsequential sounds, such as murmurs. The proposed method also uses five different classification models with 30-fold cross-validation to validate the peaks found in the noisy data. The SVM classifier achieves the highest classification accuracy and precision (94.75% and 92%, respectively), making it an efficient method for diagnosing abnormal heart sounds.

One of the potential disadvantages of the proposed approach is that it may not be as accurate as traditional heart sound classification methods that use more features to differentiate between abnormal and normal cardiac sounds. Additionally, the proposed approach relies on the use of a publicly available database, which may not accurately represent the diversity of heart sounds encountered in clinical settings. Finally, the proposed method may not be suitable for real-time applications, as it may require significant computational resources to classify heart sounds accurately.

6.6 Conclusion

In this chapter, we have introduced a novel approach for peak localization using wavelet-based K-means clustering. We demonstrated how the abovementioned algorithm is more efficient and also predicts the peaks accurately. By only using the knowledge of the detected peaks, the ML and DL models can classify the PCG signals as either normal or abnormal. The

Table 6.3 Comparison of our work with other state-of-the-art implementations.

Authors	Method	Accuracy (%)
Humanyu et al. [24]	CNN layers having tConv unit	82.27
Renna et al. [15]	Deep convolutional neural network	93.70
Xiao et al. [25]	Convolutional neural network	93.00
Gjoreski et al. [16]	Deep Learning Model Learn from Spectro-temporal portrayal	92.90
Noman et al. [23]	Markov-switching auto-regressive (MSAR)	91.20
Al-Naami et al. [17]	Adaptive neuro-fuzzy inference system (ANFIS),	89.00
Shuvo et al. [26]	CardioXNet, a new light end-to-end CRNN structure	86.57
Boulares et al. [27]	Pretrained CNN mode	89.00
Chen et al. [28]	CNN and modified frequency slice wavelet transform (MFSWT)	93.91
Li et al. [29]	CNN	86.80
S. K. Ghosh et al. [30]	TFD-DNN	95.25
	SVM	95.43
S. K. Ghosh et al. [31]	WST combined with a random forest classifier	95.12
J. Karhade et al. [32]	TFD-DL	85.16
Present work	**Wavelet-based K-means clustering with SVM**	**94.75**

proposed method not only outperforms state-of-the-art baseline systems in terms of performance, but it is also computationally cheap to categorize signals as abnormal or normal. Moreover, we have used an algorithm using an unsupervised ML model, which is faster and can automatically localize the fundamental heart sounds without calculating any features. As per our experimental reports, SVM obtained the highest accuracy (94.75%) mentioned in Table 6.2. We compared our work to other state-of-the-art works in Table 6.3 in order to validate the performance of our proposed methodology. The high-level outcomes produced by the proposed method confirm its usefulness in computer-supported heart sound examination and give a stage for applications requiring localization of murmurs and ejection clicks and a variety of subsequent examinations. We can test the effectiveness and robustness of this approach on real-time data in the future.

References

[1] S.K. Ghosh, R. Ponnalagu, R. Tripathy, U.R. Acharya, Deep layer kernel sparse representation network for the detection of heart valve ailments from the time-frequency representation of PCG recordings, BioMed Research International 2020 (2020).

[2] S.K. Ghosh, P.R. Nagarajan, R.K. Tripathy, Heart sound data acquisition and preprocessing techniques: a review, in: Handbook of Research on Advancements of Artificial Intelligence in Healthcare Engineering, 2020, pp. 244–264.

[3] G.D. Clifford, C. Liu, B.E. Moody, J.M. Roig, S.E. Schmidt, Q. Li, I. Silva, R.G. Mark, Recent advances in heart sound analysis, Physiological Measurement 38 (2017) E10–E25.

[4] S. Mangione, Cardiac auscultatory skills of physicians-in-training: a comparison of three English-speaking countries, The American Journal of Medicine 110 (3) (2001) 210–216.

[5] Amit Krishna Dwivedi, Syed Anas Imtiaz, Esther Rodriguez-Villegas, Algorithms for automatic analysis and classification of heart sounds—a systematic review, IEEE Access 7 (2018) 8316–8345.

[6] V. Arora, R. Leekha, R. Singh, I. Chana, Heart sound classification using machine learning and phonocardiogram, Modern Physics Letters B 33 (26) (2019) 1950321.

[7] V. Sujadevi, K. Soman, R. Vinayakumar, A. Prem Sankar, Anomaly detection in phonocardiogram employing deep learning, in: Computational Intelligence in Data Mining: Proceedings of the International Conference on CIDM 2017, Springer, 2019, pp. 525–534.

[8] E. Messner, M. Zöhrer, F. Pernkopf, Heart sound segmentation—an event detection approach using deep recurrent neural networks, IEEE Transactions on Biomedical Engineering 65 (9) (2018) 1964–1974.

[9] C.D. Papadaniil, L.J. Hadjileontiadis, Efficient heart sound segmentation and extraction using ensemble empirical mode decomposition and kurtosis features, IEEE Journal of Biomedical and Health Informatics 18 (4) (2013) 1138–1152.

[10] D.B. Springer, L. Tarassenko, G.D. Clifford, Logistic regression-HSMM-based heart sound segmentation, IEEE Transactions on Biomedical Engineering 63 (4) (2015) 822–832.

[11] S. Barma, B.-W. Chen, W. Ji, S. Rho, C.-H. Chou, J.-F. Wang, Detection of the third heart sound based on nonlinear signal decomposition and time–frequency localization, IEEE Transactions on Biomedical Engineering 63 (8) (2015) 1718–1727.

[12] T. Fernando, H. Ghaemmaghami, S. Denman, S. Sridharan, N. Hussain, C. Fookes, Heart sound segmentation using bidirectional LSTMs with attention, IEEE Journal of Biomedical and Health Informatics 24 (6) (2019) 1601–1609.

[13] T. Dissanayake, T. Fernando, S. Denman, S. Sridharan, H. Ghaemmaghami, C. Fookes, A robust interpretable deep learning classifier for heart anomaly detection without segmentation, IEEE Journal of Biomedical and Health Informatics 25 (6) (2020) 2162–2171.

[14] S.M. Lundberg, S.-I. Lee, A unified approach to interpreting model predictions, Advances in Neural Information Processing Systems 30 (2017).

[15] F. Renna, J. Oliveira, M.T. Coimbra, Deep convolutional neural networks for heart sound segmentation, IEEE Journal of Biomedical and Health Informatics 23 (6) (2019) 2435–2445.

[16] M. Gjoreski, A. Gradišek, B. Budna, M. Gams, G. Poglajen, Machine learning and end-to-end deep learning for the detection of chronic heart failure from heart sounds, IEEE Access 8 (2020) 20313–20324.

[17] B. Al-Naami, H. Fraihat, N.Y. Gharaibeh, A.-R.M. Al-Hinnawi, A framework classification of heart sound signals in PhysioNet Challenge 2016 using high order statistics and adaptive neuro-fuzzy inference system, IEEE Access 8 (2020) 224852–224859.

[18] T.-E. Chen, S.-I. Yang, L.-T. Ho, K.-H. Tsai, Y.-H. Chen, Y.-F. Chang, Y.-H. Lai, S.-S. Wang, Y. Tsao, C.-C. Wu, S1 and S2 heart sound recognition using deep neural networks, IEEE Transactions on Biomedical Engineering 64 (2) (2016) 372–380.

[19] K.A. Babu, B. Ramkumar, M.S. Manikandan, Automatic identification of S1 and S2 heart sounds using simultaneous PCG and PPG recordings, IEEE Sensors Journal 18 (22) (2018) 9430–9440.

[20] V.N. Varghees, K. Ramachandran, Effective heart sound segmentation and murmur classification using empirical wavelet transform and instantaneous phase for electronic stethoscope, IEEE Sensors Journal 17 (12) (2017) 3861–3872.

[21] A. Likas, N. Vlassis, J.J. Verbeek, The global K-means clustering algorithm, Pattern Recognition 36 (2) (2003) 451–461.

[22] A.L. Goldberger, L.A. Amaral, L. Glass, J.M. Hausdorff, P.C. Ivanov, R.G. Mark, J.E. Mietus, G.B. Moody, C.-K. Peng, H.E. Stanley, PhysioBank, PhysioToolkit, and PhysioNet: components of a new research resource for complex physiologic signals, Circulation 101 (23) (2000) e215–e220.

[23] F. Noman, S.-H. Salleh, C.-M. Ting, S.B. Samdin, H. Ombao, H. Hussain, A Markov-switching model approach to heart sound segmentation and classification, IEEE Journal of Biomedical and Health Informatics 24 (3) (2019) 705–716.

[24] A.I. Humayun, S. Ghaffarzadegan, M.I. Ansari, Z. Feng, T. Hasan, Towards domain invariant heart sound abnormality detection using learnable filterbanks, IEEE Journal of Biomedical and Health Informatics 24 (8) (2020) 2189–2198.

[25] B. Xiao, Y. Xu, X. Bi, J. Zhang, X. Ma, Heart sounds classification using a novel 1-D convolutional neural network with extremely low parameter consumption, Neurocomputing 392 (2020) 153–159.

[26] S.B. Shuvo, S.N. Ali, S.I. Swapnil, M.S. Al-Rakhami, A. Gumaei, CardioXNet: a novel lightweight deep learning framework for cardiovascular disease classification using heart sound recordings, IEEE Access 9 (2021) 36955–36967.

[27] M. Boulares, T. Alafif, A. Barnawi, Transfer learning benchmark for cardiovascular disease recognition, IEEE Access 8 (2020) 109475–109491.

[28] Y. Chen, S. Wei, Y. Zhang, Classification of heart sounds based on the combination of the modified frequency wavelet transform and convolutional neural network, Medical & Biological Engineering & Computing 58 (2020) 2039–2047.

[29] F. Li, H. Tang, S. Shang, K. Mathiak, F. Cong, Classification of heart sounds using convolutional neural network, Applied Sciences 10 (11) (2020) 3956.

[30] S.K. Ghosh, R.N. Ponnalagu, R.K. Tripathy, G. Panda, R.B. Pachori, Automated heart sound activity detection from PCG signal using time–frequency-domain deep neural network, IEEE Transactions on Instrumentation and Measurement 71 (2022) 1–10.

[31] S.K. Ghosh, R.K. Tripathy, R.N. Ponnalagu, R.B. Pachori, Automated detection of heart valve disorders from the PCG signal using time-frequency magnitude and phase features, IEEE Sensors Letters 3 (12) (2019) 1–4.

[32] J. Karhade, S. Dash, S.K. Ghosh, D.K. Dash, R.K. Tripathy, Time–frequency-domain deep learning framework for the automated detection of heart valve disorders using PCG signals, IEEE Transactions on Instrumentation and Measurement 71 (2022) 1–11.

CHAPTER 7

Verifying the effectiveness of a Taylor–Fourier filter bank-based PPG signal denoising approach using machine learning

José Antonio de la O Serna[a], **Rajesh Kumar Tripathy**[b], **Alejandro Zamora-Mendez**[c], **and Mario R. Arrieta Paternina**[d]

[a]Autonomous University of Nuevo Leon, Monterrey, NL, Mexico
[b]Department of EEE, Birla Institute of Technology and Science, Pilani, Hyderabad, India
[c]Michoacan University of Saint Nicholas of Hidalgo, Morelia, Michoacan, Mexico
[d]National Autonomous University of Mexico (UNAM), Mexico City, Mex., Mexico

Contents

7.1 Introduction

Photoplethysmogram (PPG) signals are contaminated with various types of artifacts during recording. A high–quality PPG signal is essential for applications such as the detection of cardiac diseases [1], blood glucose monitoring for diabetes detection [2] [3], emotion recognition [4], and human activity recognition [5]. Signal processing techniques are vital in filtering various artifacts from PPG recordings [6]. PPG signals are contaminated with low- and high–frequency artifacts [7] [3]. The motion artifact is a type of low-frequency artifact that occurs due to finger movement during the recording of the PPG signal [8] [9] [10]. Similarly, power-line interference causes high-frequency artifacts in the PPG signal [7]. Developing novel signal processing approaches for filtering artifacts from PPG signals is a challenging problem in biomedical signal processing.

In recent years, various methods have been proposed to filter low-frequency and high-frequency artifacts from PPG recordings [11] [12]. Yan et al. [13] used the smoothed pseudo-Wigner–Ville distribution (SPWVD)-based time-frequency representation (TFR) for the removal of motion artifacts from PPG data. They performed postprocessing using the PPG signal's time-frequency-domain representation to remove motion artifacts. Joseph et al. [14] used discrete wavelet transform (DWT) to decompose PPG signals into wavelet coefficients at different scales. They applied a thresholding operation over wavelet coefficients followed by DWT to eliminate high-frequency artifacts from PPG data. In [15], the authors used wavelet transform to remove motion artifacts from PPG signals. Challenges in wavelet-based techniques to remove artifacts from PPG data include the proper selection of the mother wavelet and the number of decomposition levels. The empirical mode decomposition (EMD)-based approach for removing motion artifacts has been proposed by Raghu et al. [10]. Due to the mode-mixing effect in EMD-based methods, it is challenging to select artifact-free modes or intrinsic mode functions (IMFs) to evaluate filtered PPG signals. Pollreisz and TaheriNejad [16] used a band-pass filter (BPF) to filter out motion artifacts from PPG recordings. In another study, Ram et al. [17] proposed the adaptive step-size-based least mean square (LMS) approach to eliminate motion artifacts from PPG recordings. In [18], the authors applied independent component analysis (ICA) using a two-channel PPG signal to eliminate motion artifacts. Liang et al. [19] compared the denoising performance of various filters to eliminate high-frequency artifacts (muscle artifacts) from the PPG signals. They reported that the Chebyshev II fourth-order filter effectively eliminated the muscle artifacts better than other filters. The LMS-based adaptive filter approach requires a reference PPG signal to filter out the motion artifacts [17]. Similarly, blind source separation methods, such as ICA, require multiple-channel PPG signals and independent component (IC) selection to filter artifacts. In addition, the nonreference-based approaches, such as wavelet, EMD, and BPF, did not perform well in removing artifacts from the PPG data. The existing methods have used correlation coefficients and mean square error (MSE) measures to evaluate the performance of PPG denoising algorithms. Thus, a PPG signal denoising approach with better performance and a new objective quality measure to evaluate PPG signal filtering algorithms are required.

The O-spline implemented discrete-time Taylor–Fourier transform (DTTFT) filter bank has been proposed by de la O Serna [20] for the analysis of power signals, EEG signals, and other nonstationary signals [21] [22] [23]. The filters implemented using O-splines precisely capture the frequency information at each harmonic band and allow for reconstruction of the time behavior [20]. Due to the exploitation of both frequency and time domains, the DTTFT is an ideal technique for decomposing nonstationary signals. The PPG signal is nonstationary; therefore, the O-spline-based DT-TFT filter bank has not been explored to analyze this signal. The novelty of this work lies in developing a set of finite impulse response (FIR) filters using O-spline-based DTTFT to eliminate both low- and high-frequency artifacts from the recorded PPG signals. There are variations in the time-domain and frequency-domain information of the filtered PPG signal after eliminating different artifacts. Hence, new methods with better denoising performance can be developed for the filtering of artifacts from PPG data. The major contributions of this chapter are as follows:

- The O-spline implemented DTTFT filter bank is introduced to decompose noisy PPG signals into modes or subband signals.
- A mode selection criterion based on the frequency estimation using a differentiator of O-splines is formulated, and the filtered PPG signal is reconstructed using the selected modes.
- The effectiveness of the proposed O-spline-based denoising approach is verified using two classification strategies: normal vs. HT and LA vs. HA using PPG signals.

The remaining sections of this chapter are organized as follows. In Section 7.2, we describe the details regarding the PPG signal databases used in this work. The proposed O-spline-based PPG signal filtering approach is described in Section 7.3. We show the results obtained using our method and discuss the results in Section 7.4. Finally, the conclusions of this work are mentioned in Section 7.5.

7.2 PPG signal database

In this study, we have used PPG recordings from two public databases to evaluate the performance of the proposed O-spline-based denoising approach. The first database used in this work is the PPG blood pressure (PPG-BP) database[1] [24]. This database is the collection of 720 PPG recordings obtained from 219 subjects [24]. The systolic and diastolic blood pressure values for each subject have been given in the database. The PPG signal has been recorded from the fingertip of the left index finger [24]. Subjects aged 21–86 years were considered for the recording. Each PPG signal is sampled at a sampling frequency of 1 kHz and comprises 2100 samples (duration, 2.1 min) in each PPG recording. In this database, the PPG signals are contaminated with high-frequency artifacts (muscle artifacts). We have considered 237 PPG recordings from the normal class and 231 PPG recordings from the HT class for our study.

The second database used in this study is the DEAP database[2] [25]. This database comprises various physiological signals, such as multichannel electroencephalography (EEG), electrocardiography (ECG), electromyography (EMG), electrooculography (EOG), galvanic skin response (GSR), electroretinography (ERG), and PPG signals, recorded during various emotional states when the subjects were watching music videos [25]. PPG signals in the DEAP database are contaminated with motion artifacts, and the sampling frequency of each PPG signal is 128 Hz. The PPG recordings from 32 subjects are used in this work. Each PPG recording consists of 40 trials in the second database, containing 8064 samples. The arousal score for each trail of PPG signal is given in the DEAP database [25]. PPG trials with an arousal score greater than 4.5 are considered as high arousal (HA) [25]. Similarly, an arousal score of less than 4.5 for the PPG signal is considered as low arousal (LA). The plots of motion artifact-contaminated PPG signals (recording number S01.mat) from the DEAP database and PPG signals contaminated with muscle artifacts (recording number as 58_2.txt) from the PPG-BP database are shown in Fig. 7.1(a) and Fig. 7.1(c), respectively. Similarly, the spectra of the low-frequency artifact- and high-frequency artifact-contaminated PPG signals are depicted in Fig. 7.1(b) and Fig. 7.1(d), respectively. For the motion artifact case, the amplitude value is high in the frequency range below 0.5 Hz.

7.3 Method

The proposed O-spline-based PPG signal denoising approach is shown in Fig. 7.2. It mainly consists of decomposing a PPG signal into subband sig-

---

[1] https://figshare.com/articles/dataset/PPG-BP_Database_zip/5459299.
[2] https://www.eecs.qmul.ac.uk/mmv/datasets/deap/.

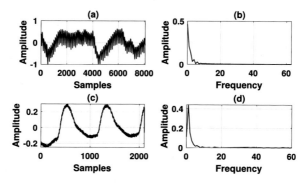

Figure 7.1 Example of waveforms and spectra for PPG signals containing motion and muscle artifacts. (a) Motion artifact-contaminated PPG signal from the DEAP database. (b) The spectrum of the motion artifact-contaminated PPG signal. (c) A high-frequency artifact (muscle artifact) contaminated the PPG signal from the PPG-BP database. (d) The spectrum of the high-frequency artifact-contaminated PPG signal.

nals or modes using the O-spline implemented Taylor–Fourier filter bank, selection of modes using the estimated frequency based on the first O-spline differentiator, evaluation of the filtered PPG signal, and verification of the diagnostic quality of PPG signal using two different classification frameworks. The following subsections describe each stage of the proposed approach.

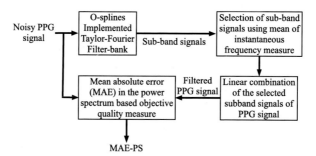

Figure 7.2 Strategy for eliminating high- and low-frequency artifacts from PPG signals using the proposed O-spline implemented Taylor–Fourier filter bank.

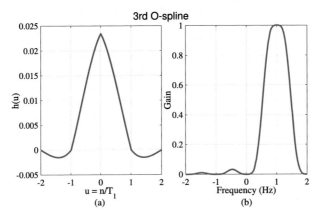

Figure 7.3 Taylor–Fourier's FIR filter implemented with the third O-spline ($K = 3$) for extracting the subband signals and their frequency. (a) Impulse response. (b) Frequency response.

7.3.1 O-spline FIR filter implementation

The discrete-time Taylor–Fourier transform is a signal decomposition approach proposed by de la O et al. [26] for analyzing nonstationary signals [27]. It has been used for the analysis of electrocardiogram (ECG) and electroencephalogram (EEG) signals [28] [22]. The low-pass O-splines and their derivatives are obtained in closed form from the theoretical solution of the DTTFT, once it is factorized into two separated matrices: one for the Taylor operator Υ and the other for the Fourier one Ω [20]. They have been found in the vectors of the dual matrix of the Taylor matrix Φ_0 obtained when Fourier matrices W_N in the Fourier operator are reduced to their first vector of ones. In this way, for a given Taylor order K, the O-spline and its first K derivatives are orthogonal to the set of $K + 1$ Taylor terms in the Taylor matrix, forming a biorthogonal pair of bases for the Taylor subspace limited by that order, because we have $\widetilde{\Phi}_0^T \Phi_0 = I$. The role of the Fourier operator is to modulate the O-spline and its derivatives at each harmonic frequency included in the sampling frequency interval, forming a set of bandpass filters that perform as ideal bandpass differentiators.

In this work, we just need to extract 16 subbands from the PPG signals and we need 16 FIR filters of $(K+1)N$ length per observed frequency, with $K = 3$. This means that the third-order O-spline represented by its impulse response in Fig. 7.3(a) is used since it represents the envelope of the DTTFT filters and exhibits maximally flat pass–bands and zero flat stop-bands for the harmonic components, as shown in Fig. 7.3(b). Thereby, its closed form is found in the first column of the dual matrix of the Taylor operator [20]. After rearranging the elements of its vertical diagonal submatrices, the dual matrix becomes

$$
\tilde{\varphi}_0^{(3)}(u) = \begin{cases} \frac{1}{6}(u+3)(u+2)(u+1) & \text{for } -2 \le u < -1, \\ -\frac{1}{2}(u+2)(u+1)(u-1) & \text{for } -1 \le u < 0, \\ \frac{1}{2}(u+1)(u-1)(u-2) & \text{for } 0 \le u < 1, \\ -\frac{1}{6}(u-1)(u-2)(u-3) & \text{for } 1 \le u < 2, \\ 0 & \text{otherwise,} \end{cases} \tag{7.1}
$$

where $u = n/T_1$, with $n = 0, 1, ..., N-1$ samples of PPG signals and $T_1 = 1/F_1$, and F_1 is the specific filter bandwidth. Due to the O-splines' intrinsic symmetry, the remaining elements of the dual Taylor matrix may be computed using only two polynomials. Thus, in this work, we just need to extract 16 subbands and their first derivatives, which are used to estimate the frequency of each subband. Therefore, the filter bank of order K, for the first subband, can be allocated in the following matrix with only two vectors:

$$
\tilde{\Phi}_1^{(K)} = \left(\tilde{\varphi}_0^{(K)}(u) e^{j2\pi u} \quad -F_1 \overset{\leftharpoonup}{\phi}_0^{(K)}(u) e^{j2\pi u} \right). \tag{7.2}
$$

These vectors contain the amplitude envelopes shared by the harmonic O-splines at each harmonic frequency. While implementing the O-splines and their derivatives as filters, the computational complexity of the DTTFT is reduced considerably, preventing the matrix inversion of the numerical solution and reducing to a small number of FIR filters during the estimation operation [20]. For the PPG analysis, 16 FIR filters as specified by Table 7.1 are designed. The parameter F_c indicates the central frequency and $[F_{min}, F_{max}]$ are the bounds for the desired pass-band. The sampling frequency used for this work is $F_s = 1$ kHz, with a sampling time T_s of 1.0 ms. It may be noted that the same filter parameters as mentioned in Table 7.1 have been used to design 16 band-pass filters to decompose the PPG signals into subband signals. A bandwidth of $F_1 = 2$ Hz is employed as a design criterion; also, the filters are designed with $N_1 = 1000$ since they have equal spacing among them.

Thus, bandpass filters of DTTFT are simply modulated versions of O-splines at a particular harmonic frequency and are denoted as $\mathbf{h}_1, \mathbf{h}_2, \cdots, \mathbf{h}_k, \cdots, \mathbf{h}_{15}$. Thereby, the ith impulse response, $\mathbf{h}_i(u), \forall i = 0, ..., 15$, is given by

$$
\mathbf{h}_i(u) = [h_{i1} \ h_{i2} \cdots h_{ik} \cdots h_{iN_1}] * e^{jFc\theta_1/N_1}, \tag{7.3}
$$

Table 7.1 Central frequency and bandwidth per filter to be designed by a Taylor–Fourier's filter bank with 16 FIR filters. Column one is for the filter number. Columns two and fourth stand for the left and right frequency limits. Column three represents the central frequency.

Filter No.	F_{min} (Hz)	F_c (Hz)	F_{max} (Hz)
1	−0.5	0	0.5
2	0.5	1	1.5
3	1.5	2	2.5
4	2.5	3	3.5
5	3.5	4	4.5
6	4.5	5	5.5
7	5.5	6	6.5
8	6.5	7	7.5
9	7.5	8	8.5
10	8.5	9	9.5
11	9.5	10	10.5
12	10.5	11	11.5
13	11.5	12	12.5
14	12.5	13	13.5
15	13.5	14	14.5
16	14.5	15	15.5

where h_{i1}, h_{i2}, $\cdots$, h_{ik}, $\cdots$, h_{iN_1} are the rows of (7.1), Fc is in compliance with Table 7.1, and $\theta_1 = 2\pi/F_s$ for the 15 impulse responses. The top plot of Fig. 7.4 depicts the frequency response of the implemented FIR filters following the O-splines' design.

In this way, the subbands are obtained using the convolution ($\star$) of the PPG signal with impulse responses of the filters obtained using O-splines as

$$\mathbf{z}_i = \mathbf{s} \star \mathbf{h}_i, \tag{7.4}$$

where $\mathbf{s}$ denotes the PPG signal or set of PPG signals.

Here, $\mathbf{z}_i = [z_i(n)]_{n=1}^{n=N}$ with $i = 0, 1, 2, 3, ..., 15$, where $z_i(n)$ is denoted as the ith subband signal (see Fig. 7.5).

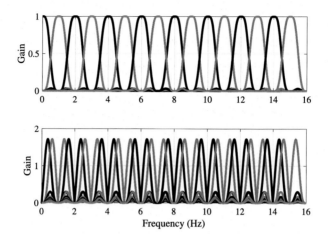

Figure 7.4 Frequency responses for filter design according to the central, minimum, and maximum frequencies in Table 7.1. (a) Sixteen FIR bandpass filters. (b) Their first-order differentiators.

7.3.2 Frequency estimation

Now, with the impulse responses of bandpass differentiators illustrated in the bottom plot of Fig. 7.4, it is possible to estimate the frequency for each subband with high accuracy. This can also converge to an ideal bandpass differentiator, as shown in the bottom plot of Fig. 7.4.

Thus, the first differentiator modulated at the ith harmonic is given by

$$\mathbf{f}_i(u) = [f_{i1} \, f_{i2} \, \cdots \, f_{ik} \, \cdots \, f_{iN_1}] * e^{jFc\theta_1/N_1}, \tag{7.5}$$

where $f_{i1}, f_{i2}, \cdots, f_{ik}, \cdots, f_{iN_1}$ are the rows of the second column of $\tilde{\varphi}_1^{(K)}$ in (7.2).

Similarly, the derivatives of the subbands are obtained by the convolution of the PPG signal and the impulse responses of bandpass differentiators. This is given by

$$\dot{\mathbf{z}}_i = \mathbf{s} * \mathbf{f}_i. \tag{7.6}$$

Therefore, the process to estimate the frequency of each subband is as follows:

1. Estimate the amplitudes of subband signals with $\hat{a}_i(n) = 2|z_i(n)|$.
2. Evaluate the phases of subband signals with $\hat{\varphi}_i(n) = \angle z_i(n)$.
3. Assess the phases of subbands' first derivatives with $\dot{\hat{\varphi}}_i(n) = 2\frac{\text{Im}\{\dot{z}_i(n)e^{-j\hat{\varphi}_i(n)}\}}{\hat{a}_i(n)}$.

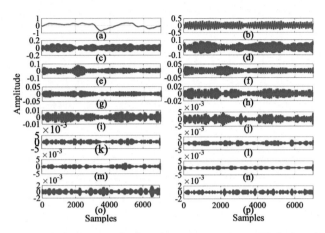

Figure 7.5 Subbands captured after decomposing one PPG signal using the first 16 FIR filters (Mode 0 to Mode 15) in Fig. 7.4.

Thus, the frequency estimation is given by

$$\hat{F}_i(n) = F_c + \frac{\hat{\varphi}_i}{2\pi}. \tag{7.7}$$

It is worth noting that no preprocessing or postprocessing stages are needed when using the dynamic model of the O-splines. Based on this analysis, 16 subband signals of the PPG signal evaluated using the O-spline filter bank are shown in Fig. 7.5(a–p). The significant spectral energy of the PPG signal is observed in the frequency range of 0.5–10 Hz [29]. The motion artifacts appear in the PPG signal below a frequency of 0.5 Hz. The zeroth subband signal, $\mathbf{z}_0$, captures the frequency range of -0.5 Hz to 0.5 Hz in the spectrum of the PPG signal. Similarly, the high-frequency noise appears above a frequency of 10 Hz. Based on the IF values computed from the subband signals using the first differentiator of the O-splines, we have selected 10 subband signals with average IF values in the frequency range of 0.5–10 Hz. The filtered PPG signal is evaluated using the linear combination of the 1st to 10th subband signals, and it is given as follows:

$$\tilde{s}(n) = \sum_{i=1}^{10} z_i(n). \tag{7.8}$$

For comparison purposes, DWT, EMD, BPF, and the median filter have been used to eliminate both low- and high-frequency artifacts from the

PPG signals. The denoising performance of the proposed O-spline filter bank is compared to that of these methods for removing both types of PPG artifacts. For DWT, we have considered the Daubechies mother wavelet ("db4") [30] and the number of decomposition levels as "7" for the PPG signals of both databases. The filtered PPG signal is evaluated using the selected subband signals from the DWT stage in the frequency range of 0.5–10 Hz. Similarly, the second-order Butterworth bandpass filter with lower and upper cutoff frequencies of 0.5 Hz and 10 Hz is used to filter PPG signals [16]. For EMD-based methods, the filtered PPG signal is evaluated based on the selection of IMFs of the contaminated PPG signal using the average value of the instantaneous frequency of each IMF within the frequency range of 0.5–10 Hz [31] [10]. For removing high-frequency artifacts from the PPG signals, we have considered the median filter of 250th order [32]. Meanwhile, an 80th-order median filter is used to eliminate low-frequency artifacts from the PPG signal. In this work, a new objective quality measure is proposed for quantifying distortion in the PPG signals after the elimination of artifacts. This measure is defined as the MAE-PS between contaminated PPG and filtered PPG signals. The MAE-PS measure (dB/Hz) for the PPG signal is given as follows:

$$\text{MAE-PS} = \frac{\sum_{f=f1}^{fN} \left| S(f) - \tilde{S}(f) \right|}{fN - f1}, \tag{7.9}$$

where $S(f)$ and $\tilde{S}(f)$ denote the PSD of the contaminated PPG signal, $s(n)$, and the filtered PPG signal, $\tilde{s}(n)$. The Welch method is used to evaluate the PSD of the PPG signal [33]. The Kaiser window is used in the Welch method for the evaluation of PSD [33]. For the first database, a window size of 512, with an overlap of 256 and 1024 FFT points, is considered for the evaluation of the PSD of the PPG signals. For the second database, a window size of 64, with an overlap of 32 and 128 FFT points, is considered. Moreover, $f1$ and fN are the lower and upper cutoff frequencies, respectively, to evaluate the MAE-PS measure for the PPG signal. In this work, the values of $f1$ and fN are considered as 0.5 Hz and 10 Hz, respectively.

7.3.3 Machine learning-based PPG quality measurement

In this work, we have also verified the effectiveness of the proposed PPG denoising approach using two different machine learning–based classification frameworks. For the first PPG signals database, the normal vs. HT

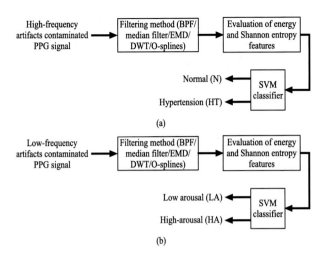

Figure 7.6 Flowchart for assessing the performance of PPG signal filtering methods using different classification schemes. (a) Normal vs. hypertension (HT). (b) Low-arousal (LA) vs. high-arousal (HA). The filtered PPG signals are evaluated using the proposed O-splines and other methods such as band-pass filter (BPF), median filter, empirical mode decomposition (EMD), and discrete wavelet transform (DWT).

classification framework is used [34] [35]. The block diagram of this classification framework is shown in Fig. 7.6(a). The original PPG signals from both normal and HT classes are fed to the proposed O-splines being compared with other filtering approaches. Then, the energy and Shannon entropy features are evaluated from the filtered PPG signals [22]. The 2D feature vector ([energy, entropy]) of the filtered PPG signal is used in the classification stage for the categorization of normal and HT classes using the support vector machine (SVM) classifier [36]. The 2D feature vectors from 237 normal and 231 HT–based PPG signals are evaluated. Hold-out validation with 70% and 30% of instances from both classes of PPG signals as training and testing sets is used to evaluate the performance of the SVM classifier.

For the second PPG signal database, the LA vs. HA emotion classification is considered. Arousal is a physiological state in which our body experiences various emotional changes based on certain stimuli [37]. The arousal scale is defined using the level of various emotional states such as aggression, excitement, stress, and nervousness [38]. Anger is interpreted as HA emotion, whereas sadness is labeled as LA emotion [37]. There are changes in the characteristics of various physiological signals, such as ECG, PPG, EEG, EMG, and GSR, during LA and HA emotional states [39]. In this study, the automated analysis of PPG signals is performed to classify high-arousal and low-arousal emotions. The block diagram for LA vs. HA classification using PPG signals is shown in Fig. 7.6(b). We have evaluated 1280 PPG signals from 32 PPG recordings in the second database. The energy and entropy features are extracted from each PPG signal, and the 2D feature vector is formulated. The SVM classifier is used to classify the 2D feature vector into LA and HA classes. We have considered 60% of PPG signals as the training set, and the remaining 40% of PPG signals are used to test the SVM classifier. The performance of the SVM classification model is evaluated using the accuracy, the Cohen kappa score, and the Matthews correlation coefficient (MCC) for both databases [40].

7.4 Results and discussion

In this section, we show the results obtained using the proposed O-spline-based DTTFT filter bank and other approaches such as DWT, EMD, BPF, and the median filter to remove high- and low-frequency artifacts from the PPG signals.

7.4.1 High-frequency noise filtering

The PPG signal polluted with high-frequency artifacts is shown in Fig. 7.7(a). The filtered signal is evaluated using BPF, median filter, EMD, and DWT, and it is compared with the proposed O-splines. All approaches are depicted in Fig. 7.7(b–f). The MAE-PS values evaluated for O-splines, DWT, EMD, BPF, and the median filter are 0.5827, 0.7275, 3.0887, 0.8202, and 1.0462 dB/Hz, respectively. The classification results obtained for different denoising algorithms with SVM classifier using 2D feature vectors of the filtered PPG signals are shown in Table 7.2. It can be observed from Table 7.2 that for 2D feature vectors obtained from original (noisy) PPG signals, the SVM classifier yielded accuracy, kappa, and MCC values of 47.14%, −0.061, and −0.063, respectively. The classification performance of SVM is lower due to the presence of high-frequency artifacts in the PPG data during recording. When the feature vectors of the filtered PPG signals are obtained using the proposed O-spline-based filter bank, the accuracy, kappa, and MCC values of the SVM classifier are found as 52.14%, 0.042, and 0.042, respectively. The feature vectors of the filtered PPG signals evaluated using other filtering techniques such as BPF, median

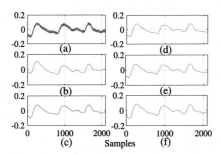

Figure 7.7 (a) Plots of PPG signals contaminated with high-frequency arti-facts. (b–f) The filtered PPG signals are obtained using (b) a band-pass filter (BPF); (c) a median filter; (d) empirical mode decomposition (EMD); (e) discrete wavelet transform (DWT); and (f) the proposed DTTFT filter bank implemented with O-splines.

filter, EMD, and DWT achieved lower classification performance using the SVM classifier as compared to the proposed O-spline filter bank-based PPG denoising approach. The proposal preserved the PPG signal infor-mation after eliminating high–frequency artifacts. Hence, the classification performance of the SVM classifier is improved using the feature vectors of the O-spline-based filtered PPG signals.

Table 7.2 Accuracy values of the SVM classifier evaluated using high-frequency artifact-contaminated PPG signals and filtered PPG signals using different approaches for the normal vs. HT classification scheme.

Method	Noisy PPG	BPF	Median filter	EMD	DWT	Proposed O-splines
Accuracy (%)	47.14	38.57	50	45.71	50.71	52.14
Kappa	−0.061	−0.226	0.001	−0.081	0.017	0.042
MCC	−0.063	−0.228	0.001	−0.085	0.017	0.042

7.4.2 Low-frequency noise filtering

The PPG signal contaminated by low-frequency artifacts is shown in Fig. 7.8, whereas the filtered PPG signals computed by using BPF, median filter, EMD, DWT, and the proposed filter bank are shown in Fig. 7.8(b–e). The MAE-PS values obtained for the filtered sig-nals are shown in Fig. 7.8(b–d); values of 2.3252, 2.1211, 1.7143, and 1.8777 dB/Hz are obtained for BPF-, median filter-, EMD-, and DWT–based methods, respectively. The O-spline filter bank exhibits better de-

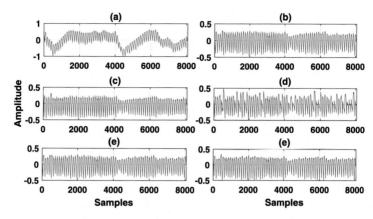

Figure 7.8 Results of denoising the PPG signals contaminated with low-frequency artifacts in (a) with the following filtering methods: (b) band-pass filter (BPF); (c) median filter; (d) empirical mode decomposition (EMD); (e) discrete wavelet transform (DWT); and (f) the DTTFT filter bank implemented with O-splines.

noising performance with the lowest MAE-PS value among all filtering techniques to eliminate low-frequency artifacts from the PPG data.

Table 7.3 Accuracy values of the SVM classifier evaluated using low-frequency artifact-contaminated PPG signals and PPG signals filtered using different approaches for the LA vs. HA classification scheme.

Method	Noisy PPG	BPF	Median filter	EMD	DWT	Proposed O-splines
Accuracy (%)	57.42	59.76	58.20	60.15	60.54	60.93
Kappa	−0.065	−0.022	−0.046	0.0002	0.002	0.015
MCC	−0.080	−0.028	−0.056	0.0003	0.003	0.018

The results obtained using the SVM classifier for the LA vs. HA classification scheme with 2D feature vectors of filtered PPG signals are shown in Table 7.3. It can be noted that we have obtained an accuracy value of 57.42% using feature vectors of noisy PPG signals. Negative kappa and MCC values are observed for noisy PPG signal feature vectors using the SVM classifier. When the feature vectors of the O-spline–based filtered PPG signals are used, the SVM classifier demonstrates higher accuracy, kappa, and MCC values. However, the feature vectors computed using DWT- and EMD-based filtered PPG signals achieve classification results close to those achieved by the proposed O-spline filter bank with SVM classifier. The proposed O-spline filter bank demonstrates higher denoising performance than other methods for filtering low- and high-frequency artifacts from PPG signals.

In this work, we have also compared the computational complexity of the denoising algorithms to filter the artifacts from PPG signals. The computational complexity of the EMD-based PPG signal decomposition algorithm is given as $O(N_s(41NN_I))$ [41]. The parameters N_s and N_I denote the number of shifting operations involved in EMD and the number of modes, respectively [41]. The factor N is denoted as the length of the PPG signal. Similarly, the computational complexity of DWT-based decomposition of PPG signal is $O(N)$ [30]. The computational complexity of the proposed O-spline-based filter bank is evaluated as $C[(N\log_2(N)/2 + 1) + 1]$ [20], where $C = K + 1$ and K is the order of the O-spline. In this work, a third-order O-spline is considered for the design of the filter bank. It has been observed that the proposed O-spline filter bank has lower computational complexity than the EMD algorithm. The DWT has the lower computational complexity as compared to the EMD and O-spline filter bank methods. Though the computational complexity of the proposed O-spline filter bank is higher than that of DWT, it performed well in eliminating both low- and high-frequency artifacts from PPG signals compared to the DWT approach. The 0–20 Hz frequency range is used to obtain the high-fidelity PPG signals [42]. In this study, we added five more filters to our proposed O-spline filter bank to extend its frequency range from 0–15 Hz to 0–20 Hz. The classification results of the SVM model are also evaluated using the filtered PPG signals obtained from a 20-filter-based O-spline filter bank for the NT vs. HT classification scheme. The accuracy, kappa, and MCC values are obtained as 52.14, 0.042, and 0.042, respectively. It is observed that the classification performance of the SVM model remains the same after using the features from the filtered PPG signals obtained using a 20-filter-based O-spline filter bank.

Moreover, we have also evaluated the classification performance of the SVM model using the morphological features such as the average of the peak amplitude values and an average of the peak values of the first derivative of the PPG signals [43,44] for the filtering of low- and high-frequency artifacts; the results are summarized in Table 7.4. The accuracy values of the SVM classifier are obtained as 52.85% and 62.10%, respectively, using the morphological features of O-spline-based filtered PPG signals to filter high- and low-frequency artifacts. The accuracy values of the SVM using features of O-spline-based filtered PPG signals are higher than those obtained using features of PPG signals evaluated using BPF and the medium filter. The proposed O-spline-based PPG signal denoising approach preserved the peaks in the PPG signal and the first derivative of the PPG signal. Due to this reason, the SVM classifier has obtained a higher classification performance using the morphological features from the O-spline-based approach compared to EMD-, DWT-, BPF-, and median filter-based PPG signal denoising approaches. The proper selection of order in BPF and the window size in the median filter are required to filter artifacts from the PPG signal [19]. However, the proposed O-spline-based filter bank does not require any tunable parameters to remove artifacts from the PPG recordings. Even though the BPF and median filter have lower computational times than the proposed O-spline filter bank, the features of the filtered PPG signals evaluated using these filters have less classification accuracy using the SVM classifier for both classification schemes.

Table 7.4 Performance of the SVM classifier using morphological features of O-spline-based filtered PPG signals.

Artifact	Accuracy (%)	Kappa	MCC
Low-frequency	62.10	0.015	0.022
High-frequency	52.85	0.057	0.058

7.4.3 Discussion

The objective of this work is the development of an O-spline-based Taylor–Fourier filter bank for the removal of low-frequency and high-frequency artifacts from PPG signals. Using features of the proposed O-spline-based filtered PPG signals, the SVM classifier has demonstrated higher average accuracy, average MCC, and average kappa scores than the features of raw PPG signals and the features of the filtered PPG signals evaluated using other denoising methods for the normal vs. HT classification task. Moreover, for the LA vs. HA classification strategy, the classification results of the SVM model are improved using features from the proposed O-spline-based filtered PPG signals compared to the features of the raw PPG signals and PPG signals filtered using other methods. The advantages of the proposed O-spline-based approach are given as follows:

- The proposed O-spline-based approach outperforms other denoising methods for filtering low-frequency and high-frequency artifacts from PPG data.

- The classification performance of the SVM model is improved after removing artifacts from the PPG signals using the proposed approach.
- The suggested approach can be used as a preprocessing step for the automated detection of pathologies using PPG signals.

The PPG signal is used in different applications such as blood pressure estimation, continuous monitoring of heart rate variability, detection of cardiac ailments, and sleep monitoring [1], [2] [3]. Filtering artifacts from this signal is a preprocessing step for accurate analysis of PPG data and detection of various pathologies. The proposed O-spline-based approach can be used as a preprocessing task for the real-time processing of PPG data in different healthcare applications. The low-frequency artifacts considered in this study are a mixture of low-frequency components and baseline wandering noise. Eliminating low-frequency artifacts is not a good idea for applications like sleep apnea detection using PPG signals. The features from the respiratory components of the PPG signal can be used to detect sleep apnea. In the future, more robust algorithms based on various deep learning methods can be developed for denoising low-frequency artifacts and extracting respiratory components from the PPG signals.

7.5 Conclusion

An approach based on the Taylor–Fourier filter bank implemented with the third-order O-spline and its first derivative has been proposed in this work to eliminate low-frequency and high-frequency artifacts from PPG recordings. The PPG signals have been decomposed into subband signals using the proposed O-spline-based filter bank. The average value of the instantaneous frequency measure is used to select subband signals to evaluate the filtered PPG signal. The robustness of the proposed denoising approach has been verified through classification frameworks such as normal vs. HT and LA vs. HA using PPG signals. The classification accuracy of the SVM classifier has been improved after eliminating various artifacts from the PPG signals, demonstrating the superiority of our proposed O-spline-based method.

References

[1] T. Pereira, N. Tran, K. Gadhoumi, M.M. Pelter, D.H. Do, R.J. Lee, R. Colorado, K. Meisel, X. Hu, Photoplethysmography based atrial fibrillation detection: a review, npj Digital Medicine 3 (1) (2020) 1–12.

[2] G. Zhang, Z. Mei, Y. Zhang, X. Ma, B. Lo, D. Chen, Y. Zhang, A noninvasive blood glucose monitoring system based on smartphone PPG signal processing and machine learning, IEEE Transactions on Industrial Informatics 16 (11) (2020) 7209–7218.

[3] Samuel Huthart, Mohamed Elgendi, Dingchang Zheng, Gerard Stansby, John Allen, Advancing PPG signal quality and know-how through knowledge translation—from experts to student and researcher, Frontiers in Digital Health (ISSN 2673-253X) 2 (2020), https://doi.org/10.3389/fdgth.2020.619692.

[4] A. Goshvarpour, A. Goshvarpour, Poincaré's section analysis for PPG-based automatic emotion recognition, Chaos, Solitons and Fractals 114 (2018) 400–407.

[5] M. Boukhechba, L. Cai, C. Wu, L.E. Barnes, ActiPPG: using deep neural networks for activity recognition from wrist-worn photoplethysmography (PPG) sensors, Smart Health 14 (2019) 100082.

[6] K. Xu, X. Jiang, S. Lin, C. Dai, W. Chen, Stochastic modeling based nonlinear Bayesian filtering for photoplethysmography denoising in wearable devices, IEEE Transactions on Industrial Informatics 16 (11) (2020) 7219–7230.

[7] M. Elgendi, On the analysis of fingertip photoplethysmogram signals, Current Cardiology Reviews 8 (1) (2012) 14–25.

[8] D. Seok, S. Lee, M. Kim, J. Cho, C. Kim, Motion artifact removal techniques for wearable EEG and PPG sensor systems, Frontiers in Electronics (2021) 4.

[9] C.-H. Goh, L.K. Tan, N.H. Lovell, S.-C. Ng, M.P. Tan, E. Lim, Robust PPG motion artifact detection using a 1-D convolution neural network, Computer Methods and Programs in Biomedicine 196 (2020) 105596.

[10] M. Raghuram, K.V. Madhav, E.H. Krishna, N.R. Komalla, K. Sivani, K.A. Reddy, HHT based signal decomposition for reduction of motion artifacts in photoplethysmographic signals, in: 2012 IEEE International Instrumentation and Measurement Technology Conference Proceedings, IEEE, 2012, pp. 1730–1734.

[11] B. Roy, R. Gupta, MoDTRAP: improved heart rate tracking and preprocessing of motion-corrupted photoplethysmographic data for personalized healthcare, Biomedical Signal Processing and Control 56 (2020) 101676.

[12] M.S. Islam, M. Shifat-E-Rabbi, A.M.A. Dobaie, M.K. Hasan, PREHEAT: precision heart rate monitoring from intense motion artifact corrupted PPG signals using constrained RLS and wavelets, Biomedical Signal Processing and Control 38 (2017) 212–223.

[13] Y.-s. Yan, C.C. Poon, Y.-t. Zhang, Reduction of motion artifact in pulse oximetry by smoothed pseudo Wigner-Ville distribution, Journal of NeuroEngineering and Rehabilitation 2 (1) (2005) 1–9.

[14] G. Joseph, A. Joseph, G. Titus, R.M. Thomas, D. Jose, Photoplethysmogram (PPG) signal analysis and wavelet de-noising, in: 2014 Annual International Conference on Emerging Research Areas: Magnetics, Machines and Drives (AICERA/iCMMD), IEEE, 2014, pp. 1–5.

[15] C. Lee, Y.T. Zhang, Reduction of motion artifacts from photoplethysmographic recordings using a wavelet denoising approach, in: IEEE EMBS Asian-Pacific Conference on Biomedical Engineering, 2003, IEEE, 2003, pp. 194–195.

[16] D. Pollreisz, N. TaheriNejad, Detection and removal of motion artifacts in PPG signals, Mobile Networks and Applications (2019) 1–11.

[17] M.R. Ram, K.V. Madhav, E.H. Krishna, N.R. Komalla, K.A. Reddy, A novel approach for motion artifact reduction in PPG signals based on AS-LMS adaptive filter, IEEE Transactions on Instrumentation and Measurement 61 (5) (2011) 1445–1457.

[18] B.S. Kim, S.K. Yoo, Motion artifact reduction in photoplethysmography using independent component analysis, IEEE Transactions on Biomedical Engineering 53 (3) (2006) 566–568.

[19] Y. Liang, M. Elgendi, Z. Chen, R. Ward, An optimal filter for short photoplethysmogram signals, Scientific Data 5 (1) (2018) 1–12.

[20] J.A. de la O Serna, Dynamic harmonic analysis with FIR filters designed with O-splines, IEEE Transactions on Circuits and Systems 67 (12) (2020) 5092–5100.

[21] D. Guillen, J.A. de la O Serna, A. Zamora-Méndez, M.R.A. Paternina, F. Salinas, Taylor–Fourier filter-bank implemented with O-splines for the detection and classification of faults, IEEE Transactions on Industrial Informatics 17 (5) (2020) 3079–3089.

[22] J.A. de la O Serna, M.R.A. Paternina, A. Zamora-Méndez, R.K. Tripathy, R.B. Pachori, EEG-rhythm specific Taylor–Fourier filter bank implemented with O-splines for the detection of epilepsy using EEG signals, IEEE Sensors Journal 20 (12) (2020) 6542–6551.

[23] J.A. de la O Serna, Analyzing power oscillating signals with the O-splines of the discrete Taylor–Fourier transform, IEEE Transactions on Power Systems 33 (6) (2018) 7087–7095.

[24] Y. Liang, Z. Chen, G. Liu, M. Elgendi, A new, short-recorded photoplethysmogram dataset for blood pressure monitoring in China, Scientific Data 5 (1) (2018) 1–7.

[25] S. Koelstra, C. Muhl, M. Soleymani, J.-S. Lee, A. Yazdani, T. Ebrahimi, T. Pun, A. Nijholt, I. Patras, DEAP: a database for emotion analysis; using physiological signals, IEEE Transactions on Affective Computing 3 (1) (2011) 18–31.

[26] J.A. de la O Serna, J.M. Ramirez, A.Z. Mendez, M.R.A. Paternina, Identification of electromechanical modes based on the digital Taylor-Fourier transform, IEEE Transactions on Power Systems 31 (1) (2015) 206–215.

[27] M.A. Platas-Garza, J.A. de la O Serna, Dynamic harmonic analysis through Taylor–Fourier transform, IEEE Transactions on Instrumentation and Measurement 60 (3) (2010) 804–813.

[28] R.K. Tripathy, A. Zamora-Mendez, J.A. de la O Serna, M.R.A. Paternina, J.G. Arrieta, G.R. Naik, Detection of life threatening ventricular arrhythmia using digital Taylor Fourier transform, Frontiers in Physiology 9 (2018) 722.

[29] N. Sviridova, K. Sakai, Human photoplethysmogram: new insight into chaotic characteristics, Chaos, Solitons and Fractals 77 (2015) 53–63.

[30] P. Gajbhiye, N. Mingchinda, W. Chen, S.C. Mukhopadhyay, T. Wilaiprasitporn, R.K. Tripathy, Wavelet domain optimized Savitzky–Golay filter for the removal of motion artifacts from EEG recordings, IEEE Transactions on Instrumentation and Measurement 70 (2020) 1–11.

[31] N.E. Huang, Z. Shen, S.R. Long, M.C. Wu, H.H. Shih, Q. Zheng, N.-C. Yen, C.C. Tung, H.H. Liu, The empirical mode decomposition and the Hilbert spectrum for nonlinear and non-stationary time series analysis, Proceedings of the Royal Society of London. Series A: Mathematical, Physical and Engineering Sciences 454 (1971) (1998) 903–995.

[32] J. Lee, Motion artifacts reduction from PPG using cyclic moving average filter, Technology and Health Care 22 (3) (2014) 409–417.

[33] D.-J. Jwo, W.-Y. Chang, I.-H. Wu, Windowing techniques, the Welch method for improvement of power spectrum estimation, Computers, Materials & Continua 67 (2021) 3983–4003, https://doi.org/10.32604/cmc.2021.014752.

[34] V.R. Nafisi, M. Shahabi, Intradialytic hypotension related episodes identification based on the most effective features of photoplethysmography signal, Computer Methods and Programs in Biomedicine 157 (2018) 1–9.

[35] K.-c. Lan, P. Raknim, W.-F. Kao, J.-H. Huang, Toward hypertension prediction based on PPG-derived HRV signals: a feasibility study, Journal of Medical Systems 42 (6) (2018) 1–7.

[36] C. Cortes, V. Vapnik, Support-vector networks, Machine Learning 20 (3) (1995) 273–297.

[37] D. Derryberry, M.K. Rothbart, Arousal, affect, and attention as components of temperament, Journal of Personality and Social Psychology 55 (6) (1988) 958.

[38] D. Maheshwari, S. Ghosh, R. Tripathy, M. Sharma, U.R. Acharya, Automated accurate emotion recognition system using rhythm-specific deep convolutional neural network technique with multi-channel EEG signals, Computers in Biology and Medicine 134 (2021) 104428.

[39] G. Chanel, C. Rebetez, M. Bétrancourt, T. Pun, Emotion assessment from physiological signals for adaptation of game difficulty, IEEE Transactions on Systems, Man and Cybernetics. Part A. Systems and Humans 41 (6) (2011) 1052–1063.

[40] M. Sokolova, G. Lapalme, A systematic analysis of performance measures for classification tasks, Information Processing & Management 45 (4) (2009) 427–437.

[41] Y.-H. Wang, C.-H. Yeh, H.-W.V. Young, K. Hu, M.-T. Lo, On the computational complexity of the empirical mode decomposition algorithm, Physica A: Statistical Mechanics and Its Applications 400 (2014) 159–167.

[42] H. Liu, J. Allen, S.G. Khalid, F. Chen, D. Zheng, Filtering-induced time shifts in photoplethysmography pulse features measured at different body sites: the importance of filter definition and standardization, Physiological Measurement 42 (7) (2021) 074001, https://doi.org/10.1088/1361-6579/ac0a34.

[43] S. Vadrevu, M.S. Manikandan, Real-time PPG signal quality assessment system for improving battery life and false alarms, IEEE Transactions on Circuits and Systems. II, Express Briefs 66 (11) (2019) 1910–1914.

[44] G.N.K. Reddy, M.S. Manikandan, N.N. Murty, Evaluation of objective distortion measures for automatic quality assessment of processed PPG signals for real-time health monitoring devices, IEEE Access 10 (2022) 15707–15745.

CHAPTER 8

Automated detection of hypertension from PPG signals using continuous wavelet transform and transfer learning

Shresth Gupta, Anurag Singh, and Abhishek Sharma
Department of ECE, IIIT Naya Raipur, Naya Raipur, Chhattisgarh, India

Contents

8.1 Introduction

The death rate of people as a result of cardiovascular diseases (CVDs) has increased significantly [1]. The World Health Organization (WHO) released figures showing that the death rate from CVD will increase from 246 per million in 2015 to 264 per million in 2030 [2]. It is well recognized that having abnormally high blood pressure can impair your heart, kidneys, and other essential organs permanently. Therefore, the prevention and treatment of CVDs may greatly benefit from early hypertension diagnosis, medication, and management. In this regard, photoplethysmography (PPG) is a feasible and efficient technology that can be used to monitor hypertension and a variety of cardiovascular indicators, such as blood oxygen saturation, heart rate, cuffless blood pressure, diabetes, mean arterial pressure, respiration rate, and arterial aging [3–7]. By using PPG, we potentially measure volumetric changes in the blood inside an organ or throughout the body.

Electrocardiography (ECG) and PPG have been utilized extensively with machine learning (ML) models for blood pressure estimation. Both the estimate of blood pressure and the categorization of the stages of hypertension have been employed using these models in blood pressure research. Liang et al. used a feature collectively obtained from ECG and PPG called pulse arrival time and also extracted several morphological features of PPG and classified the blood pressure categories [8]. Lopez et al. [9] proposed a logistic regression model to determine if a person has hypertension. Their model's sensitivity is 77%, the specificity is 68%, and the area under the receiver operating characteristic curve (AUC) is 0.73. Patnaik et al. [10] examined the effectiveness of a number of classification techniques, including naive Bayes (NB) classifiers, support vector machine (SVM), logistic regression models, random forest (RF), and multilayer perceptron (MLP). With an AUC of 0.8977 and an accuracy of 80.23%, the SVM model achieved the highest degree of accuracy. A work utilized the first and second derivatives of the PPG signal and its Hilbert–Huang transform (HHT) and used the CNN model for classification, getting F1-scores of 98.90%, 85.80%, and 93.54%, respectively, for three different classes of hypertension [11]. The problems associated with manual feature extraction include the accurate detection of feature points from each PPG cycle. Further, the approach also requires optimal filtering before feature extraction to remove unwanted noisy components that create barriers in key-point detection. The limitations are resolved by the proposed framework for automatic diagnosis of hypertension by converting the second derivative of a short recorded PPG episode into its corresponding scalogram (by using the continuous wavelet transform). The obtained images are fed to the transfer learning module to obtain the multiclass classification of hypertension. The proposed work depicts three broad classes of hypertension, namely, normal, prehypertension, and hypertension.

8.2 Methodology

The block diagram shown in Fig. 8.1 illustrates the pipeline of the proposed work. The preprocessing block performs the necessary optimal filtering of the raw PPG corrupted with noise, followed by a block that produces the

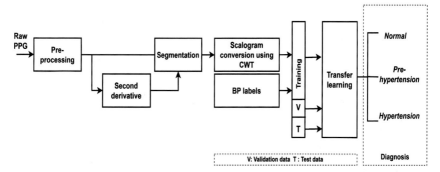

Figure 8.1 End-to-end framework for the diagnosis of hypertension using transfer learning.

second derivative of the clean PPG, also called acceleration plethysmography (APG). This 1D time series APG is converted into 2D scalogram images. The scalogram images are given to the transfer learning block that comprises pretrained deep learning models that finally provide classification in three broad hypertension classes. The detailed methodology is provided in the below subsections.

8.2.1 Database

The 657 PPG records from 219 people in the publicly available PPG-BP database are used in this work [12]. The database contains information on illnesses including hypertension at different stages, as well as information on patients from 20 to 89 years old. The expected test settings were used when collecting the data. PPG signals are recorded at a frequency of 1 kHz. The database also includes other significant vital parameters and demographic data of subjects, such as systolic and diastolic blood pressure, heart rate, body mass index, age, sex, etc., in order to create algorithms to forecast various cardiovascular diseases and brain abnormalities. The statistics of this database are illustrated in Fig. 8.2, which highlights the histogram of the subjects' age groups and the classwise portion of available data.

8.2.2 Preprocessing of raw PPG records

In order to remove outliers and associated artifacts from the raw PPG data, we experimented with a few popular filtering approaches used in the previous literature, including Savitzky–Golay filtering [13], wavelet-based denoising [14], and the Chebyshev-II filtering approach [15]. Amongst all

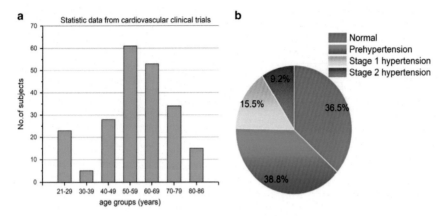

Figure 8.2 Statistical information of the database used in the proposed work [12]. (a) Histogram of age groups of the subjects. (b) Percentage of available subjects in each blood pressure class.

these, Chebyshev-II filtering performed better due to the reason that, along with a finer transition zone, the Chebyshev-II filter also features a flat, ripple-free passband, while the stopband has a similar amount of ripple. As little as possible interference with the signal's usable component in the passband attributed to these qualities is caused. As can be seen in Fig. 8.3, which consists of three short concatenated noisy PPG records having high-frequency noise and morphological deviations. Chebyshev-II can filter out this interference and noise while retaining the useful information of the signal because it has great frequency selectivity and no equal ripple in the pass band. The obtained PPG indicates the corrected shape with no high-frequency components in the contour, which facilitates the detection of fiducial points.

8.2.3 Continuous wavelet transform

Continuous wavelet transform (CWT) is a signal processing technique used to analyze and extract information from signals that vary over time. Some of the key advantages of CWT include [16]:

- Time–frequency localization: CWT provides excellent time–frequency localization, allowing for the analysis of signals that have both high-frequency and low-frequency components that vary over time. This is because CWT uses a family of wavelet functions that are scaled and shifted to match the frequency content of the signal at different points in time.

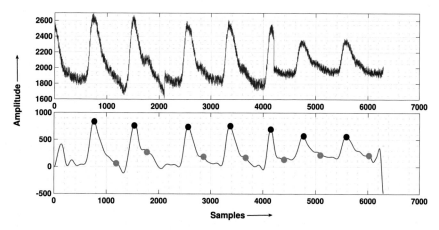

Figure 8.3 Denoising of PPG signal using the Chebyshev-II filtering approach. The detected systolic and diastolic peaks are marked in the clean PPG.

- Multiresolution analysis: CWT can analyze signals at multiple resolutions, allowing for the identification of both fine and coarse features in the signal. This is because CWT uses wavelets of different scales, which provide different levels of detail in the analysis.
- Flexibility: CWT can be adapted to suit different applications and signal types. This is because the choice of wavelet function and other parameters can be adjusted to optimize the analysis for a specific signal or application.
- Data compression: CWT can be used to compress signals by removing redundant or unnecessary information while preserving important features. This is because the multiresolution analysis provided by CWT allows for the identification of features that are important for a specific application.
- Interpretability: CWT provides interpretable results that can be used to gain insights into the characteristics of the signal being analyzed. This is because the time-frequency localization and multiresolution analysis provided by CWT can reveal important features of the signal that are not apparent in the raw data.

The term "wavelet analysis function" usually refers to a function family made up of the scaling function and the shrinking function $\psi(t)$. The function $\psi(t)$ must satisfy the following requirement [17]:

$$\int_{-\infty}^{+\infty} \psi(t)\,dt = 0. \tag{8.1}$$

Then, $\psi_{p,q}(t)$ can be deduced as follows:

$$\psi_{p,q}(t) = \frac{1}{\sqrt{|p|}} \psi\left(\frac{t-q}{p}\right), \quad p, q \in R, \quad p \neq 0, \qquad (8.2)$$

where the base wavelet (mother wavelet function) is $\psi(t)$ and p and q are the associated dilation (scaling) and translation (shrinking) parameters. The mother wavelet-based wavelet function is described in Eq. (8.2). Assuming that $\psi_{p,q}$ is the wavelet function for Eq. (8.2), for any function specified in L2 space, a continuous wavelet transfer is defined as

$$Wf(a, b) = \frac{1}{\sqrt{a}} \int_{-\infty}^{+\infty} f(t)\psi^*\left(\frac{t-b}{a}\right) dt. \qquad (8.3)$$

The wavelet space is affected by the values of p and q in addition to the wavelet analysis window function. This special characteristic makes the wavelet transfer an adaptable technique for decomposing nonlinear signals. CWT analytic domains usually shift to the research aspects.

8.2.4 PPG-to-scalogram conversion

The CWT of a signal's absolute value, shown as a function of time and frequency, is known as a scalogram. Scalograms can reveal the signal's low-frequency and fast-evolving frequency components [17]. The PPG signals are 1D vector signals which should be recognized. The 1D temporal segment PPG is transformed into a 30-second PPG scalogram with a "jet" colormap. The CWT was used to create an RGB picture while maintaining the nonstationary signals' time-frequency localization properties. In comparison to the short-time Fourier transform, scalograms provide greater time localization for short-duration, high-frequency events and better frequency localization for low-frequency, long-duration events. (See Fig. 8.4.)

8.2.5 Transfer learning approach

Transfer learning is the term used in ML to describe using a previously trained model on a different task. In transfer learning, a computer leverages the information gained from previous work to improve prediction about a new task. We used two transfer learning models, namely, DenseNet201 and VGG19, for hypertension diagnosis [18]. The architecture of these pre-trained deep learning models is illustrated in Fig. 8.5.

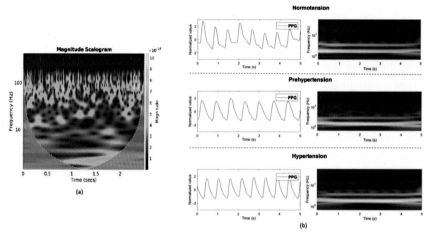

Figure 8.4 (a) Demonstration of a standard magnitude scalogram (jet-128 bit) indicating the time-frequency response of a signal. (b) PPG signals of different blood pressure categories and their corresponding scalogram images.

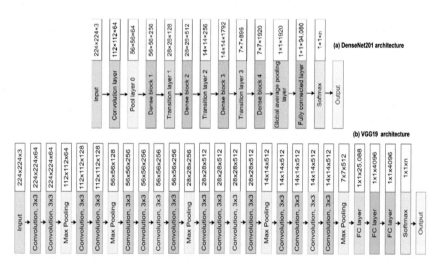

Figure 8.5 Architecture of pretrained deep networks with allowed dimension in each layer. (a) DenseNet201. (b) VGG-19.

8.2.5.1 DenseNet201 architecture

DenseNet is based on the belief that convolutional networks may be trained to be significantly deeper, more precise, and more effective if the connections between the layers near the input and the layers near the output are shorter. Each layer is linked to every other layer in a DenseNet topology, giving rise to the term densely connected convolutional network. There are $L(L+1)/2$ direct connections for L layers (Fig. 8.5(a)). The feature maps of all the layers before it are used as inputs for each layer, and its own feature maps are utilized as inputs for each layer after it [19]. DenseNets simply connect every layer to every other layer. The main, incredibly potent notion is this. A layer in DenseNet receives its input as a concatenation of feature maps from earlier levels. DenseNets offer a variety of convincing benefits, including the elimination of the vanishing-gradient issue, improved feature propagation, promoted feature reuse, and much fewer parameters.

8.2.5.2 VGG19 architecture

Simonyan and Zisserman (2014) introduced VGG19, a convolutional neural network of 19 layers, 16 convolution layers, and 3 fully connected layers that can categorize images into 1000 different object categories [20]. The ImageNet database, which has a million photos in 1000 categories, served as the training data for VGG19. Due to the employment of numerous 3×3 filters in each convolutional layer, it is a highly well-liked technique for classifying images. Fig. 8.5(b) depicts the architecture of VGG19. This demonstrates that the first 16 convolutional layers are used to extract features, while the next 3 layers are utilized to do classification. The layers used for feature extraction are divided into five groups, with a max-pooling layer after each group. This model receives an image with a size of 224 by 224 and produces the label of the entity in the picture. In the article, features are retrieved using a pretrained VGG19 model, while other ML approaches are used for classification. The size of the feature vector must be reduced by dimensionality reduction since the CNN model computes a large number of parameters after feature extraction. With locality-preserving projection (LPP), the dimensionality is reduced, and then a classification approach is used.

8.3 Results

The available short recorded PPG signals are first grouped into three major hypertension classes, i.e., normal, prehypertension, and hypertension, in which the hypertension class contains both stage-I and stage-II hypertension categories. For each record, we have also obtained its second derivative (also called APPG). Using APPG along with natural PPG has two advantages in our proposed approach:

Table 8.1 Details of optimal hyperparameters used in the pretrained model of the proposed diagnosis framework.

Hyperparameter	Value	Hyperparameter	Value
First layer input size	$224 \times 224 \times 3$	Max epochs	8
Initial learn rate	0.0001	Mini batch size	12
Learn rate drop period	10	Validation frequency	5
Learn rate drop factor	0.1	Validation patience	60
L2 regularization	0.0001	Verbose frequency	50
Momentum	0.9	Gradient decay factor	0.95

- The use of APPG elevates the appearance of most of the useful fiducial points that may not be visible in their natural form [21]. The converted scalogram from these signals can extract the clinical information in the time frequency more accurately and increase the diagnosis performance.

- As we have limited records for each class of hypertension, the use of both PPG and APPG doubles the available records that yield sufficient scalogram images to train the deep model.

We converted each record in a scalogram using the color map jet-128, which provided clear RGB contours in the scalogram, indicating the time and frequency variation of the PPG signal. As per the requirement of the network, we resized the obtained scalogram in the dimension of $224 \times 224 \times 3$ for the first input layer of DenseNet and VGG19. A total of 1298 scalogram images are generated: 334 images for hypertension, 470 images for normal, and 494 images for prehypertension. We split the data for training, validation, and testing at a ratio of 60:20:20. After the initial experiment, and we used the optimal hyperparameters with both neural networks mentioned in Table 8.1.

After training and testing with both pretrained models, the best results are obtained with VGG19. The obtained performance with these two models is illustrated in Table 8.2. We can observe from the results that VGG19 performed better in the hypertension diagnosis, with an overall validation and test accuracy of 96.9% and 94%, respectively. The variation of accuracy and loss with increasing epochs is illustrated in Fig. 8.6 for the best-performing pretrained model (VGG-19). As in transfer learning, which employs a pretrained model, the training takes less time as compared to the newly designed neural network architecture.

The confusion matrix obtained after classification using the VGG-19 model for validation and test data is illustrated in Fig. 8.7. It can be ob-

Table 8.2 Classwise and overall diagnosis performance of the two pretrained models used in the proposed approach. NOR, normal; PRE, prehypertension; HYP, hypertension.

Model	No. of parameters (in millions)	Classwise accuracy (in %)			Validation accuracy (in %)	Test accuracy (in %)	Training time (in sec)
		NOR	PRE	HYP			
DenseNet201	20	89.4	100	91	94	93.8	2271
VGG-19	144	92.6	100	98.5	96.9	94	428

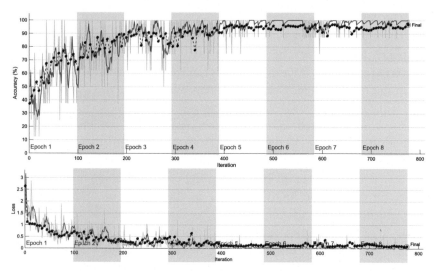

Figure 8.6 Variation of validation accuracy and loss with increasing epoch for VGG-19. The black dots represent the validation values, while the colored curves represent training accuracy and loss.

served that the prehypertension class is well diagnosed by the proposed framework with a 100% correctly classified count in validation and a single misclassification in test data. This result clearly indicates that the proposed work is suitable for early diagnosis of hypertension (as prehypertension). The time utilized by the model is also important when the model is deployed in an edge device. Thus, we have also calculated the time taken by each model to yield final classification results. Along with higher accuracy, the VGG-19 model took significantly less time (428 sec) as compared with DenseNet, which took 2271 sec.

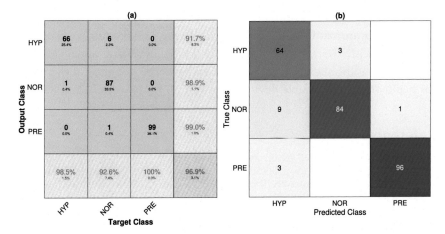

Figure 8.7 Confusion matrix indicating true and false classified counts of the three classes after classification with the VGG-19 model. (a) Validation data. (b) Test data.

8.4 Discussion

According to the results presented in the above section, the proposed approach eliminates feature engineering for PPG applied in the temporal or frequency domain for the diagnosis of hypertension. In the manual feature extraction approach, an additional module is always required to detect single or multiple fiducial points from each PPG cycle to extract morphological features. This is typically challenging when the acquired PPG has moderate or high distortions when acquired during nonresting measurement conditions. However, we have used a preprocessing approach which is always required when dealing with raw PPG data, to improve the signal-to-noise ratio and the clinical information of PPG, which is further reflected in its transformed scalogram.

8.5 Conclusion

A complete PPG-based end-to-end framework for the early diagnosis of hypertension is presented in this work with higher diagnostic accuracy. The key advantage and robustness of the proposed work lie in the use of a fiducial point-independent approach, and thus, instead of manual feature calculation, we exploited the use of CWT to capture the time-frequency information of each PPG episode in the form of scalograms. In addition, we have utilized APPG also, which carries more valuable information as compared to normal PPG. The scalogram image data are further classified using the transfer learning approach with optimal hyperparameters. Higher accuracy in both validation and testing highlights the potential of the work for deployment in wearable devices.

References

[1] WHO, A global brief on hypertension: silent killer, global public health crisis, https://www.who.int/publications/i/item/a-global-brief-on-hypertension-silent-killer-global-public-health-crisis-world-health-day-2013, 2013, WHO REFERENCE NUMBER: WHO/DCO/WHD/2013.2.

[2] P. Jeemon, T. Séverin, C. Amodeo, D. Balabanova, N.R. Campbell, D. Gaita, K. Kario, T. Khan, R. Melifonwu, A. Moran, et al., World heart federation roadmap for hypertension–a 2021 update, Global Heart 16 (1) (2021).

[3] S. Gupta, A. Singh, A. Sharma, Dynamic large artery stiffness index for cuffless blood pressure estimation, IEEE Sensors Letters 6 (3) (2022) 1–4.

[4] S. Gupta, A. Singh, A. Sharma, R.K. Tripathy, Higher order derivative-based integrated model for cuff-less blood pressure estimation and stratification using PPG signals, IEEE Sensors Journal 22 (22) (2022) 22030–22039.

[5] S. Gupta, A. Singh, A. Sharma, R.K. Tripathy, dSVRI: a PPG-based novel feature for early diagnosis of type-II diabetes mellitus, IEEE Sensors Letters 6 (9) (2022) 1–4.

[6] S. Gupta, A. Singh, A. Sharma, Exploiting moving slope features of PPG derivatives for estimation of mean arterial pressure, Biomedical Engineering Letters 13 (1) (2023) 1–9.

[7] M.A. Almarshad, M.S. Islam, S. Al-Ahmadi, A.S. BaHammam, Diagnostic features and potential applications of PPG signal in healthcare: a systematic review, Healthcare 10 (2022) 547, MDPI.

[8] Y. Liang, Z. Chen, R. Ward, M. Elgendi, Hypertension assessment via ECG and PPG signals: an evaluation using mimic database, Diagnostics 8 (3) (2018) 65.

[9] F. López-Martínez, A. Schwarcz.MD, E.R. Núñez-Valdez, V. García-Díaz, Machine learning classification analysis for a hypertensive population as a function of several risk factors, Expert Systems with Applications 110 (2018) 206–215.

[10] R. Patnaik, M. Chandran, S.-C. Lee, A. Gupta, C. Kim, C. Kim, Predicting the occurrence of essential hypertension using annual health records, in: 2018 Second International Conference on Advances in Electronics, Computers and Communications (ICAECC), 2018, pp. 1–5.

[11] X. Sun, L. Zhou, S. Chang, Z. Liu, Using CNN and HHT to predict blood pressure level based on photoplethysmography and its derivatives, Biosensors 11 (2021).

[12] Y. Liang, Z. Chen, G. Liu, M. Elgendi, A new, short-recorded photoplethysmogram dataset for blood pressure monitoring in China, Scientific Data 5 (1) (2018) 1–7.

[13] S. Gupta, A. Singh, A. Sharma, Denoising and analysis of PPG acquired from different body sites using Savitzky Golay filter, in: TENCON 2022-2022 IEEE Region 10 Conference (TENCON), IEEE, 2022, pp. 1–4.

[14] S. Gupta, A. Singh, A. Sharma, Photoplethysmogram based mean arterial pressure estimation using LSTM, in: 2021 8th International Conference on Signal Processing and Integrated Networks (SPIN), 2021, pp. 806–811.

[15] Y. Liang, M. Elgendi, Z. Chen, R. Ward, An optimal filter for short photoplethysmogram signals, Scientific Data 5 (2018).

[16] L. Aguiar-Conraria, M.J. Soares, The continuous wavelet transform: moving beyond uni- and bivariate analysis, Journal of Economic Surveys 28 (2) (2014) 344–375.

[17] John S. Sadowsky, The continuous wavelet transform: a tool for signal investigation and understanding, https://api.semanticscholar.org/CorpusID:20782920, 1994.

[18] S. Gupta, A. Gupta, A. Kumar, S. Gupta, A. Singh, Multi-class classification of colorectal cancer tissues using pre-trained CNN models, in: TENCON 2022-2022 IEEE Region 10 Conference (TENCON), IEEE, 2022, pp. 1–6.

[19] Faisal Dharma Adhinata, Diovianto Putra Rakhmadani, Merlinda Wibowo, Akhmad Jayadi, A deep learning using DenseNet201 to detect masked or non-masked face, JUITA: Jurnal Informatika 9 (2021) 115, https://api.semanticscholar.org/CorpusID: 236376662.

[20] K. Simonyan, A. Zisserman, Very deep convolutional networks for large-scale image recognition, CoRR, arXiv:1409.1556 [abs], 2014.

[21] K. Takazawa, N. Tanaka, M. Fujita, O. Matsuoka, T. Saiki, M. Aikawa, S. Tamura, C. Ibukiyama, Assessment of vasoactive agents and vascular aging by the second derivative of photoplethysmogram waveform, Hypertension 32 (2) (1998) 365–370.

CHAPTER 9

Automated estimation of blood pressure using PPG recordings: an updated review

Haipeng Liu
Centre for Intelligent Healthcare, Coventry University, Coventry, United Kingdom

Contents

9.1 Introduction

Blood pressure (BP), defined as the pressure of circulating blood against the walls of blood vessels, is a vital sign that reflects the basic hemodynamic status of the human body. The systolic and diastolic BP values (SBP and DBP, respectively) indicate the maximum and minimum pressures in a cardiac cycle. The mean arterial pressure (MAP), also called mean blood pressure (MBP) in some studies, is calculated as MAP = DBP + 1/3 (SBP − DBP). This estimate can be inaccurate in irregular cardiac rhythms, e.g., during exercise, where tachycardia shortens the diastole more than the systole. In addition, MAP is the product of two hemodynamic components: cardiac output, i.e., the flow of blood pumped by the heart each minute, and systemic vascular resistance, i.e., the total peripheral resistance [1].

Signal Processing Driven Machine Learning Techniques for
Cardiovascular Data Processing
https://doi.org/10.1016/B978-0-44-314141-6.00014-1

135

Clinically, there are three categories of BP: hypertensive, normotensive, and hypotensive. A common definition of hypertension is SBP $\geq$ 140 mmHg and/or DBP $\geq$ 90 mmHg, while hypotension is often defined as SBP < 90 mmHg and DBP < 60 mmHg [2]. Hypertension can significantly increase the risk of multiple cardiovascular conditions, e.g., stroke and myocardial infarction, and is the leading cause of cardiovascular disease and premature death worldwide [3]. Accurate estimation of BP plays a key role in the diagnosis, intervention, and management of hypertension.

The auscultatory method based on mercury sphygmomanometry has been the reference standard for office BP measurement for several decades [4]. However, it depends on manual operation. In addition, the pressure on the arm during measurement makes cuff-based methods difficult for healthcare monitoring during sleep. In recent years, noninvasive techniques for cuffless BP monitoring have demonstrated substantial advances, with photoplethysmography (PPG) playing an important role. PPG signals reflect volumetric changes in the distal vascular bed, which are dependent on BP [5]. Multiple technical approaches have been proposed for PPG-based BP estimation, where the recent development of artificial intelligence (AI) offers new possibilities for automated, cuffless, and long-term BP estimation based on wearable PPG sensors for daily healthcare.

In this chapter, the main approaches of PPG-based BP estimation are introduced regarding the theoretical basis, state-of-the-art, and application scenarios. We summarize the advantages and limitations of different methods and discuss potential future research directions. This chapter serves as a reference for biomedical engineers and healthcare professionals.

9.2 Technical approaches and challenges

PPG signals contain rich information on hemodynamics and its neural regulation. Currently, the extraction of BP from PPG signals includes four mainstream technical approaches with different physiological mechanisms. (See Fig. 9.1.)

9.2.1 Pulse transit time

Pulse transit time (PTT) refers to the time it takes a pulse wave to travel between two places in the cardiovascular system. The blood flow generates a pressure pulse wave propagating from the heart down the arterial tree through the artery walls, with a certain speed called pulse wave velocity

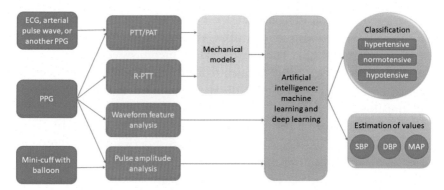

Figure 9.1 Different technical approaches of PPG-based BP estimation. ECG, electrocardiogram; PPG, photoplethysmography; PTT, pulse transit time; PAT, pulse arrival time; R-PTT, reflective PTT; SBP, systolic blood pressure; DBP, diastolic blood pressure; MAP, mean arterial pressure.

(PWV). PWV can be quantitatively evaluated by the well-known [6]:

$$PWV = \sqrt{\frac{Eh}{\rho D}}, \qquad (9.1)$$

where E is the incremental elastic modulus, h is the vessel wall thickness, D is the vessel lumen diameter, and ρ is the blood density. PTT is reciprocally related with PWV, and can be expressed by the following ratio:

$$PTT = \frac{L}{PWV}, \qquad (9.2)$$

where L is the distance between two sampled points on an arterial segment. The elastic modulus E is exponentially correlated with the mean distending pressure P:

$$E = E_0 e^{\gamma P}, \qquad (9.3)$$

where E_0 is the zero-pressure modulus, P is arterial pressure, and γ is a constant typically between 0.016 and 0.018 [6]. Therefore, the arterial pressure P can be estimated from PTT as

$$P = \frac{1}{\gamma}\left(-2\ln PTT + \ln \frac{\rho L^2 D}{hE_0}\right). \qquad (9.4)$$

With initial calibration, the SBP and DBP values can be derived.

In clinical practice, PPG-based PTT estimation can be achieved by using a reference electrocardiogram (ECG) signal or two PPG signals from different body sites [7,8]. The signals will be preprocessed to remove the baseline wandering (i.e., low-frequency fluctuations) and high-frequency noises. The ECG-PPG method is the most common one for PPG-based BP estimation, where the pulse arrival time (PAT) is used as a substitute of PTT from the heart to a distal site. PAT is defined as the time delay from the R-peak of the ECG signal to a peak of the first derivative of a PPG signal in the same cardiac cycle. Therefore, PAT includes not only the PTT, but also the time delay between electrical depolarization of the heart's left ventricle and ejection of blood through the aorta, known as the preejection period (PEP) [9]. The PEP is a main factor that affects the accuracy of PAT-derived BP estimation. Balmer et al. performed the measurement of PAT on porcine hearts and concluded that PAT is an unreliable measure of PTT and a poor surrogate under clinical interventions common in a critical care setting, due to intra- and intersubject variability in PEP [10]. A large-scale study on 2309 patients showed that PAT-derived DBP estimation can satisfy international standards for automatic oscillometric BP monitors while the estimation of SBP was not satisfactory [11].

The estimation of PTT from dual-PPG signals, i.e., two PPG probes at different body sites, could eliminate the confounding effect of PEP on the PTT results [8]. The PPG probes can be located at two sites from proximal and distal areas of an arterial segment or at different artery segments. One or both PPG signals can be substituted by an arterial pulse wave measured with other techniques, e.g., piezoelectric sensors. Samartkit et al. used a piezoelectric sensor and a PPG sensor in the calculation of PTT [12]. Dual-PPG signals from nearby sites of an artery segment provide local PWV values and enable the estimation of BP using a signal sensor, while high-quality PPG waveforms are essential for accurate estimation of PWV from the short interval [13].

An often neglected factor that affects the accuracy of PTT estimation in both ECG-PPG and dual-PPG approaches is the difference between the artery pulse waveform and the PPG waveform [14]. The calculation of PTT is based on the assumption that the pulse wave measured by a PPG sensor is the same as the pulse wave in the artery. Liu et al. pointed out that delay and deformation of PPG signals occurs compared with the pulse wave in the nearest artery [15]. Firstly, a PPG signal reflects volumetric changes, while a pulse wave reflects the pressure fluctuations. Secondly, between the artery and microcirculation where PPG signals are from, the viscoelasticity

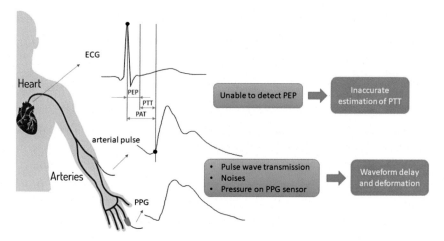

Figure 9.2 The factors that influence the accuracy of PTT methods. The illustration of a human body is adapted from [14]. PEP, preejection period.

of surrounding tissues can change the pulse wave and generate a delay. Finally, the PPG waveform is sensitive to the pressure on the sensor. (See Fig. 9.2.)

9.2.2 Reflective PTT

Reflective PTT (R-PTT) is defined as the propagation time interval between the forward and reflected pulse waves [16]. R-PTT can be extracted using a wearable PPG sensor by computing the time duration between the first PPG peak (i.e., the percussion) and the second one (reflected) in a cardiac pulsation cycle. In common PPG devices (wristband or smartwatch), physiologically, it is the time of the pulse wave propagating from the radial artery (where the wristband is located) to the end of the limb and reflecting back to the radial artery [17]. Samimi and Dajani proposed mathematical relationships between SBP, DBP, and R-PTT [18]:

$$SBP = DBP + \frac{K_a}{R - PTT^2}, \tag{9.5}$$

$$DBP = K_b + \frac{2}{0.031} \ln\left(\frac{K_c}{R - PTT}\right) - \frac{K_a}{3(R - PTT)^2}, \tag{9.6}$$

where K_a, K_b, and K_c are parameters to be calibrated for an individual in each experiment.

Compared with traditional PTT, R-PTT is more dependent on the quality of the PPG waveform. Long and Chung designed a new PPG system that can get 12 PPG signals from three different sites on the radial artery, where R-PTT and its standard deviation were used for selecting the optimal wavelength to achieve the highest-quality signal [19]. To improve the PPG signal quality for R-PTT estimation, Kao et al. integrated a critical high-order band-pass filter into the front-end circuit of the PPG sensor, where the low-pass and high-pass parts were designed for reducing ambient lighting noise and low-frequency drifts caused by breathing and/or subject motion [20]. Overall, limited by the quality of PPG signals, accurate measurement of R-PTT from PPG signals is still challenging, with a gap in validation.

9.2.3 PPG waveform analysis

The PPG waveform includes many features in different dimensions that contain rich physiological information of the cardiovascular system and can be used in BP estimation (Fig. 9.3). The measured PPG signals are typically processed with filtering, segmentation, outlier removal, and normalization for the extraction of waveform features in each cardiac cycle. Different feature points can be identified using the extrema on the velocity PPG (VPG, or the first derivative of PPG) and the acceleration PPG (APG, the second derivative of PPG) signals [21]. The areas under the PPG signal between two feature points can be calculated from the integration of the PPG signal. The normalized PPG pulse (of unit amplitude and duration) can be decomposed into the summation of four Gaussians to simplify the representation of the PPG pulse and the analysis of reflected wave interactions which are thought to have a Gaussian profile [22]. Furthermore, the PPG signal of multiple cardiac cycles can be processed with Fourier transform or wavelet transform to generate the time-frequency spectra, from which deep features can be extracted [23]. The quantitative PPG features can be inputted into machine learning algorithms for the estimation of BP values and classification of BP categories, which provides the possibility of continuous BP estimation using a single wearable sensor.

Many pathological, psychological, and medical conditions can significantly change PPG waveforms, e.g., anesthesia [24] and mental stress [25,26]. It is accepted that the changes in peripheral pulse waveform characteristics occur with aging. Allen et al. measured PPG waveforms at the ear, finger, and toe on 304 normal healthy human subjects (range 6–87

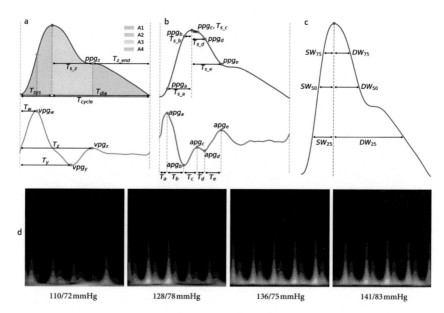

Figure 9.3 PPG waveform features for BP estimation. (a) Extrema (local maxima and minima) on VPG with corresponding periods and areas under the PPG curve. (b) Extrema of APG and periods. (c) The width of systolic and diastolic phases (SW & DW) at a given elevation of the systolic peak. (d) The time-frequency spectra of PPG signals with different BP values. (a), (b), and (c), adapted from [21]; (d), adapted from [23]. VPG, velocity PPG; APG, acceleration PPG.

years; 156 male and 148 female) and found positive correlations of the increase in PPG pulse with age (ears, fingers, and toes: +0.8, +1.9, and +1.1 ms/year, respectively), SBP (+0.5, +1.3, and + 0.9 ms/mmHg), and height (+0.5, +1.2, and +1.0 ms/cm), but a clear inverse association with heart rate (−1.8, −2.5, and −1.6 ms·min) [27]. These factors can be confounders in BP estimation across different ages. Therefore, comprehensive consideration of PPG features in a physiological context is essential to improve the reliability of BP estimation.

9.2.4 PPG with minicuff and amplitude analysis

In cuff-based methods, BP values are estimated from the modulated pulse waves under external pressure changes. The combination of PPG and miniaturized cuff can be used to observe the modulated pulse wave [28]. Bui et al. proposed an in-ear device for continuous BP monitoring us-

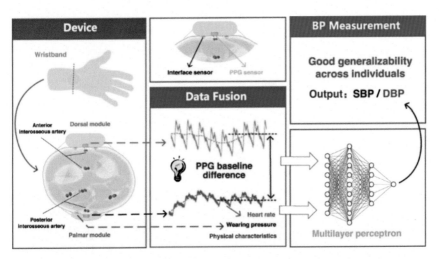

Figure 9.4 A PPG device with pressure sensor to estimate BP using machine learning. Reprint from [31].

ing a light-based inflatable pulse sensor with a PPG sensor and a digital air pump [29]. The pulse amplitude increases when the cuff pressure is close to the systolic level. The increment increases more quickly when the pressure reaches and passes through the systolic point. At the systolic and diastolic cycle cross-section, the amplitude obtains its highest value. On the other hand, the DBP position occurs at the highest decreasing amplitude. The vascular unloading technique provides a similar approach. A minicuff equipped with a PPG unit is placed around a finger. A feedback system provides pulsatile counter pressure to the cuff in order to keep the optical blood flow signal constant and thus the cuff pressure equals the intraarterial pressure [30]. Wang et al. developed a wearable device with two PPG sensors on the palmar and dorsal sides of the wrist to detect the cardiac output and the pulse waveform features with interface sensors to detect the wearing pressure and skin temperature. The detected multichannel signals were fused using a machine learning algorithm to estimate continuous BP in a real-time mode for wearers [31]. (See Fig. 9.4.)

9.3 Technical issues associated with accurate BP estimation using PPG signals

Currently, there is a lack of standardization in acquiring and processing PPG signals with a diversity in sensor design [32]. The unstable signal

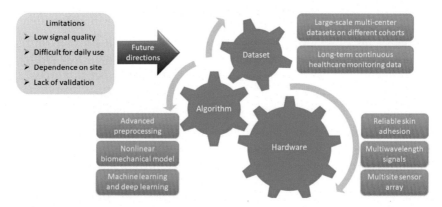

Figure 9.5 Major limitations of current approaches for PPG-based BP estimation and future research directions.

quality and the dependence on measurement site and manual operation compromise the accuracy of PPG-based BP estimation and limit its application in daily healthcare monitoring. As a result, there is a scarcity in large-scale datasets of long-term monitoring for the development and validation of AI-assisted algorithms. To address the challenges and research gaps, researchers are making various efforts to improve the hardware and algorithms (Fig. 9.5).

9.3.1 Artificial intelligence in PPG-based automatic estimation of BP

AI-assisted algorithms are reshaping the ecosystem of PPG-based BP estimation. Firstly, AI algorithms including unsupervised/semisupervised machine learning and deep learning can achieve automatic quality assessment of PPG signals [33–36]. By excluding the low-quality segments in PPG recordings, AI-assisted quality assessment can improve the performance of BP estimation in daily application scenarios where the quality of PPG signals is unstable. Besides the AI-enhanced screening, nonlinear dynamic analysis [37] and advanced biomechanical models [38,39] can improve the accuracy of BP estimation on a theoretical level. Secondly, some machine learning algorithms, including support vector machine (SVM), regression tree (RT), adaptive boosting (AdaBoost), and artificial neural network (ANN), have been adopted in PPG-based BP estimation with satisfactory performance based on a large number of data in the databases of Queensland and/or MIMIC II [40]. Machine learning has been widely used in all aforementioned technical approaches and is essential for the BP estimation based on PPG waveform features [41]. Finally, deep learning-based analysis of time-frequency spectrograms can reduce the complexity of pre-processing, which can deform PPG signals [42,43]. Pankaj et al. proposed a deep neural network to classify time-frequency spectrograms of PPG signals calculated by the Fourier decomposition method. The last three layers of the pretrained deep neural network, namely, GoogleNet, DenseNet, and AlexNet, were modified and used to classify the PPG signal into normotension, prehypertension, and hypertension with a test accuracy of 96.5% [44].

9.3.2 Skin attachment and contact pressure in reliable sensor placement

The quality and waveform of PPG signals depend on the sensor placement [45]. The stability of skin attachment is a major concern in reducing motion artifacts. In PTT-based BP estimation, the placement of multiple sensors leads to high sensitivity to motion artifacts and greatly restricts the patients' flexibility of motion [46]. Inappropriate skin attachment could lead to the complete loss of the AC component of the PPG signal [47]. Liu et al. developed a wireless, skin-interfaced device for pediatric critical care where BP was measured using a PAT-based method [48]. A soft fabric Velcro wrap with optional use of the hydrogel adheres and maintains the alignment and positioning of the ECG and PPG sensors where natural body movement could also be monitored. Another concern is that an excessive contact pressure on the skin may be transferred through the subcutaneous tissues to the microcirculation to deform the PPG signal [47]. The pressure-induced deformation of PPG signals can significantly influence the accuracy of BP estimation [49]. These issues need to be considered in the optimization of PPG sensor design.

9.3.3 Multisite PPG measurement

Block et al. [50] compared six PTTs derived from the ECG R-wave and ear PPG foot, R-wave and finger PPG foot, R-wave and toe PPG foot, ear and finger PPG feet, ear and toe PPG feet, and finger and toe PPG feet. They measured the six PTTs on 32 subjects (50% female; 52 ± 17 years, 25% hypertensive) and concluded that the PTT defined by the R-wave and toe PPG foot (toe PAT) outperformed other PTTs as a marker of changes in BP and SBP in particular [50]. Carek et al. developed a PPG sensor array at the foot to track BP using a weighing scale-based system [51]. The multisite

PPG signals can improve the robustness of BP estimation against local noises based on data selection and fusion.

9.3.4 Multiwavelength PPG signals

The depth of detection of a PPG sensor is associated with the wavelength of the light. Therefore, a multiwavelength PPG sensor can enable the detection of different anatomic structures under the skin [52]. Liu et al. developed a multiwavelength PPG sensor on the finger to measure the pulse delay through the arterioles and the capillaries, i.e., arteriole PTT, which was found to be correlated to BP [53]. Karolcik et al. developed a multisite, multiwavelength PPG platform for noninvasive monitoring of BP and other physiological parameters (heart rate, oxygen saturation, and hematocrit ratio), which enables continuous healthcare monitoring whilst further validation is still needed [54].

9.4 Conclusion

As a promising wearable sensing technology, AI-enhanced PPG signal processing and data analysis provide the possibility of automatic cuffless BP measurement in different technical approaches. Currently, some sensors and algorithms have achieved high performance, while unstable signal quality in daily monitoring and the lack of systematic validation limit its application in real-world scenarios. Future endeavors in hardware design, algorithm development, and large-scale data acquisition will further improve the reliability of PPG-based BP estimation and support its integration into the modern healthcare ecosystem.

References

[1] H.O. Ventura, S.J. Taler, J.E. Strobeck, Hypertension as a hemodynamic disease: the role of impedance cardiography in diagnostic, prognostic, and therapeutic decision making, American Journal of Hypertension 18 (S2) (2005) 26S–43S.
[2] S.G. Khalid, H. Liu, T. Zia, J. Zhang, F. Chen, D. Zheng, Cuffless blood pressure estimation using single channel photoplethysmography: a two-step method, IEEE Access 8 (2020) 58146–58154.
[3] K.T. Mills, A. Stefanescu, J. He, The global epidemiology of hypertension, Nature Reviews Nephrology 16 (4) (2020) 223–237.
[4] P. Muntner, D. Shimbo, R.M. Carey, J.B. Charleston, T. Gaillard, S. Misra, et al., Measurement of blood pressure in humans: a scientific statement from the American heart association, Hypertension 73 (5) (2019) e35–e66.
[5] A.S. Karavaev, A.S. Borovik, E.I. Borovkova, E.A. Orlova, M.A. Simonyan, V.I. Ponomarenko, et al., Low-frequency component of photoplethysmogram reflects the autonomic control of blood pressure, Biophysical Journal 120 (13) (2021) 2657–2664.

[6] X. Ding, Y.-T. Zhang, Pulse transit time technique for cuffless unobtrusive blood pressure measurement: from theory to algorithm, Biomedical Engineering Letters 9 (1) (2019) 37–52.

[7] Y. Wang, Z. Liu, S. Ma, Cuff-less blood pressure measurement from dual-channel photoplethysmographic signals via peripheral pulse transit time with singular spectrum analysis, Physiological Measurement 39 (2) (2018) 025010.

[8] G. Chan, R. Cooper, M. Hosanee, K. Welykholowa, P.A. Kyriacou, D. Zheng, et al., Multi-site photoplethysmography technology for blood pressure assessment: challenges and recommendations, Journal of Clinical Medicine 8 (11) (2019) 1827.

[9] E. Finnegan, S. Davidson, M. Harford, J. Jorge, P. Watkinson, D. Young, et al., Pulse arrival time as a surrogate of blood pressure, Scientific Reports 11 (1) (2021) 22767.

[10] J. Balmer, C. Pretty, S. Davidson, T. Desaive, S. Kamoi, A. Pironet, et al., Pre-ejection period, the reason why the electrocardiogram Q-wave is an unreliable indicator of pulse wave initialization, Physiological Measurement 39 (9) (2018) 095005.

[11] J. Lee, S. Yang, S. Lee, H.C. Kim, Analysis of pulse arrival time as an indicator of blood pressure in a large surgical biosignal database: recommendations for developing ubiquitous blood pressure monitoring methods, Journal of Clinical Medicine [Internet] 8 (11) (2019) 1773.

[12] P. Samartkit, S. Pullteap, O. Bernal, A non-invasive heart rate and blood pressure monitoring system using piezoelectric and photoplethysmographic sensors, Measurement 196 (2022) 111211.

[13] P.M. Nabeel, J. Jayaraj, S. Mohanasankar, Single-source PPG-based local pulse wave velocity measurement: a potential cuffless blood pressure estimation technique, Physiological Measurement 38 (12) (2017) 2122.

[14] G. Martínez, N. Howard, D. Abbott, K. Lim, R. Ward, M. Elgendi, Can photoplethysmography replace arterial blood pressure in the assessment of blood pressure?, Journal of Clinical Medicine [Internet] 7 (10) (2018) 316.

[15] J. Liu, B.P.-Y. Yan, W.-X. Dai, X-R. Ding, Y.-T. Zhang, N. Zhao, Multi-wavelength photoplethysmography method for skin arterial pulse extraction, Biomedical Optics Express 7 (10) (2016) 4313–4326.

[16] T.-J. Tseng, C.-H. Tseng, Noncontact wrist pulse waveform detection using 24-GHz continuous-wave radar sensor for blood pressure estimation, in: 2020 IEEE/MTT-S International Microwave Symposium (IMS), IEEE, Los Angeles, CA, USA, 2020.

[17] Y.-H. Kao, P.C.-P. Chao, C.-L. Wey, Design and validation of a new PPG module to acquire high-quality physiological signals for high-accuracy biomedical sensing, IEEE Journal of Selected Topics in Quantum Electronics 25 (1) (2018) 1–10.

[18] H. Samimi, H.R. Dajani, Cuffless blood pressure estimation using calibrated cardiovascular dynamics in the photoplethysmogram, Bioengineering [Internet] 9 (9) (2022) 446.

[19] N.M.H. Long, W.-Y. Chung, Wearable wrist photoplethysmography for optimal monitoring of vital signs: a unified perspective on pulse waveforms, IEEE Photonics Journal 14 (2) (2022) 1–17.

[20] Y.-H. Kao, P.C.P. Chao, C.-L. Wey, Towards maximizing the sensing accuracy of an cuffless, optical blood pressure sensor using a high-order front-end filter, Microsystem Technologies 24 (11) (2018) 4621–4630.

[21] S. González, W.-T. Hsieh, T.P.-C. Chen, A benchmark for machine-learning based non-invasive blood pressure estimation using photoplethysmogram, Scientific Data 10 (1) (2023) 149.

[22] E. Finnegan, S. Davidson, M. Harford, P. Watkinson, L. Tarassenko, M. Villarroel, Features from the photoplethysmogram and the electrocardiogram for estimating changes in blood pressure, Scientific Reports 13 (1) (2023) 986.

[23] J. Wu, H. Liang, C. Ding, X. Huang, J. Huang, Q. Peng, Improving the accuracy in classification of blood pressure from photoplethysmography using continuous wavelet transform and deep learning, International Journal of Hypertension 2021 (2021) 9938584.

[24] S.G. Khalid, S.M. Ali, H. Liu, A.G. Qurashi, U. Ali, Photoplethysmography temporal marker-based machine learning classifier for anesthesia drug detection, Medical & Biological Engineering & Computing 60 (11) (2022) 3057–3068.

[25] R. Udhayakumar, S. Rahman, D. Buxi, V.G. Macefield, T. Dawood, N. Mellor, et al., Measurement of stress-induced sympathetic nervous activity using multi-wavelength PPG, Royal Society Open Science 10 (8) (2023) 221382.

[26] P.H. Charlton, P. Celka, B. Farukh, P. Chowienczyk, J. Alastruey, Assessing mental stress from the photoplethysmogram: a numerical study, Physiological Measurement 39 (5) (2018) 054001.

[27] J. Allen, J. O'Sullivan, G. Stansby, A. Murray, Age-related changes in pulse rise-time measured by multi-site photoplethysmography, Physiological Measurement 41 (7) (2020) 074001.

[28] A. Chandrasekhar, C.-S. Kim, M. Naji, K. Natarajan, J.-O. Hahn, R. Mukkamala, Smartphone-based blood pressure monitoring via the oscillometric finger-pressing method, Science Translational Medicine 10 (431) (2018) eaap8674.

[29] N. Bui, N. Pham, J.J. Barnitz, Z. Zou, P. Nguyen, H. Truong, et al., eBP: an ear-worn device for frequent and comfortable blood pressure monitoring, Communications of the ACM 64 (8) (2021) 118–125.

[30] T. Panula, J.-P. Sirkiä, D. Wong, M. Kaisti, Advances in non-invasive blood pressure measurement techniques, IEEE Reviews in Biomedical Engineering 16 (2022) 424–438.

[31] L. Wang, S. Tian, R. Zhu, A new method of continuous blood pressure monitoring using multichannel sensing signals on the wrist, Microsystems & Nanoengineering 9 (1) (2023) 117.

[32] P.H. Charlton, K. Pilt, P.A. Kyriacou, Establishing best practices in photoplethysmography signal acquisition and processing, Physiological Measurement 43 (5) (2022) 050301.

[33] H. Shin, Deep convolutional neural network-based signal quality assessment for photoplethysmogram, Computers in Biology and Medicine 145 (2022) 105430.

[34] M. Feli, I. Azimi, A. Anzanpour, A.M. Rahmani, P. Liljeberg, An energy-efficient semi-supervised approach for on-device photoplethysmogram signal quality assessment, Smart Health 28 (2023) 100390.

[35] W.-K. Beh, Y.-C. Yang, Y.-C. Lo, Y.-C. Lee, A.-Y. Wu, Machine-aided PPG signal quality assessment (SQA) for multi-mode physiological signal monitoring, ACM Transactions on Computing for Healthcare 4 (2) (2023) 14.

[36] M.S. Roy, R. Gupta, K.D. Sharma, Photoplethysmogram signal quality evaluation by unsupervised learning approach, in: 2020 IEEE Applied Signal Processing Conference (ASPCON), IEEE, Kolkata, India, 2020.

[37] C. Landry, S.D. Peterson, A. Arami, Nonlinear dynamic modeling of blood pressure waveform: towards an accurate cuffless monitoring system, IEEE Sensors Journal 20 (10) (2020) 5368–5378.

[38] G. Rovas, V. Bikia, N. Stergiopulos, Quantification of the phenomena affecting reflective arterial photoplethysmography, Bioengineering [Internet] 10 (4) (2023) 460.

[39] D. Barvik, M. Cerny, M. Penhaker, N. Noury, Noninvasive continuous blood pressure estimation from pulse transit time: a review of the calibration models, IEEE Reviews in Biomedical Engineering 15 (2021) 138–151.

[40] P.C.-P. Chao, C.-C. Wu, D.H. Nguyen, B.-S. Nguyen, P.-C. Huang, V.-H. Le, The machine learnings leading the cuffless PPG blood pressure sensors into the next stage, IEEE Sensors Journal 21 (11) (2021) 12498–12510.

[41] C. El-Hajj, P.A. Kyriacou, A review of machine learning techniques in photoplethysmography for the non-invasive cuff-less measurement of blood pressure, Biomedical Signal Processing and Control 58 (2020) 101870.

[42] J. Allen, H. Liu, S. Iqbal, D. Zheng, G. Stansby, Deep learning-based photoplethysmography classification for peripheral arterial disease detection: a proof-of-concept study, Physiological Measurement 42 (5) (2021) 054002.

[43] S. Liao, H. Liu, W.-H. Lin, D. Zheng, F. Chen, Filtering-induced changes of pulse transmit time across different ages: a neglected concern in photoplethysmography-based cuffless blood pressure measurement, Frontiers in Physiology 14 (2023) 1172150.

[44] Pankaj, A. Kumar, M. Kumar, R. Komaragiri, Optimized deep neural network models for blood pressure classification using Fourier analysis-based time–frequency spectrogram of photoplethysmography signal, Biomedical Engineering Letters 13 (2023) 739–750.

[45] V. Hartmann, H. Liu, F. Chen, Q. Qiu, S. Hughes, D. Zheng, Quantitative comparison of photoplethysmographic waveform characteristics: effect of measurement site, Frontiers in Physiology 10 (2019) 198.

[46] D. Castaneda, A. Esparza, M. Ghamari, C. Soltanpur, H. Nazeran, A review on wearable photoplethysmography sensors and their potential future applications in health care, International Journal of Biosensors & Bioelectronics 4 (4) (2018) 195.

[47] F. Scardulla, G. Cosoli, S. Spinsante, A. Poli, G. Iadarola, R. Pernice, et al., Photoplethysmograhic sensors, potential and limitations: is it time for regulation? A comprehensive review, Measurement 218 (2023) 113150.

[48] C. Liu, J.-T. Kim, S.S. Kwak, A. Hourlier-Fargette, R. Avila, J. Vogl, et al., Wireless, skin-interfaced devices for pediatric critical care: application to continuous, non-invasive blood pressure monitoring, Advanced Healthcare Materials 10 (17) (2021) 2100383.

[49] A. Chandrasekhar, M. Yavarimanesh, K. Natarajan, J.-O. Hahn, R. Mukkamala, PPG sensor contact pressure should be taken into account for cuff-less blood pressure measurement, IEEE Transactions on Biomedical Engineering 67 (11) (2020) 3134–3140.

[50] R.C. Block, M. Yavarimanesh, K. Natarajan, A. Carek, A. Mousavi, A. Chandrasekhar, et al., Conventional pulse transit times as markers of blood pressure changes in humans, Scientific Reports 10 (1) (2020) 16373.

[51] A.M. Carek, H. Jung, O.T. Inan, A reflective photoplethysmogram array and channel selection algorithm for weighing scale based blood pressure measurement, IEEE Sensors Journal 20 (7) (2019) 3849–3858.

[52] S. Han, D. Roh, J. Park, H. Shin, Design of multi-wavelength optical sensor module for depth-dependent photoplethysmography, Sensors [Internet] 19 (24) (2019) 5441.

[53] J. Liu, B.P. Yan, Y.-T. Zhang, X.-R. Ding, P. Su, N. Zhao, Multi-wavelength photoplethysmography enabling continuous blood pressure measurement with compact wearable electronics, IEEE Transactions on Biomedical Engineering 66 (6) (2018) 1514–1525.

[54] S. Karolcik, D.K. Ming, S. Yacoub, A.H. Holmes, P. Georgiou, A multi-site, multi-wavelength PPG platform for continuous non-invasive health monitoring in hospital settings, IEEE Transactions on Biomedical Circuits and Systems 17 (2023) 349–361.

CHAPTER 10

Time-frequency-domain deep representation learning for detection of heart valve diseases using PCG recordings for IoT-based smart healthcare applications

Hari Krishna Damodaran[a], **Rajesh Kumar Tripathy**[a], **and Ram Bilas Pachori**[b]

[a]Department of EEE, Birla Institute of Technology and Science, Pilani, Hyderabad, India
[b]Department of Electrical Engineering, Indian Institute of Technology Indore, Indore, India

Contents

10.1 Introduction

The Internet of Things (IoT) has enabled the development of innovative solutions to improve patient care in healthcare settings [1]. The essential features of IoT for healthcare are the use of sensors to record physiological data, the transmission of patient data to the cloud, preprocessing of the physiological data in the cloud, the development of AI-based models for clinical decision making, data security, and telemedicine by connecting patients with healthcare professionals [2]. Cardiac auscultation is a diagnostic framework using a digital stethoscope to record the phonocardiography (PCG) signal produced during the cardiac cycle [3]. The PCG signal mainly contains different heart sound components, such as the first heart sound (S1), the second heart sound (S2), heart murmurs, the third heart sound (S3), and the fourth heart sound (S4) [4]. The S3- and S4-sounds appear in PCG during congestive heart failure and in the case of hypertrophic cardiomyopathy-based heart diseases [5]. However, in the case of pregnant females and children, the S3-sound is considered normal [6]. Typically, the S1- and S2-sounds have an overlapping frequency content, with the frequency ranges for these two sounds being found at approximately 20–150 Hz and 50–100 Hz, respectively [7]. The frequency ranges of systolic and diastolic murmurs are higher than those of S1 and S2 heart sound components in PCG signals [7]. The automated classification of HVDs using AI-based techniques is essential for IoT-based intelligent healthcare systems for monitoring patients' cardiac health from remote locations [1]. The PCG data recorded using a digital stethoscope can be given as input to the AI-based model, which is deployed on the cloud to detect HVDs. Developing novel AI-based algorithms and deploying these algorithms on cloud-based frameworks is essential for designing IoT-based cardiac health monitoring systems to detect HVDs.

Various methods based on machine learning and deep learning techniques have recently been reported to detect different HVDs using PCG data [8]. In [9], the authors employed the wavelet-domain synchrosqueezing transform followed by the extraction of features from the PCG data. They considered the random forest (RF) classifier to detect aortic stenosis (AS), mitral regurgitation (MR), and mitral stenosis (MS). Similarly, methods based on chirplet transform (CT)- and spline chirplet transform (SCT)-domain machine learning techniques have been used to detect HVDs using PCG data [4] [10]. Likewise, in [8], the authors extracted geometrical features in the Fourier–Bessel-domain empirical wavelet-based multiscale decomposition of PCG data and used different supervised learning techniques to detect HVDs. The graph pattern-based generation model, which uses PCG data as input, followed by the machine learning approach, has been utilized to detect HVDs [11]. In [12], the authors extracted wavelet entropy-based features from the PCG data and formulated a threshold-based decision-making approach to classify normal and abnormal classes. Ortiz et al. [13] extracted the dynamic time wrapping (DTW)- and male frequency cepstral coefficient (MFCC)-based features from the PCG data and employed a support vector machine (SVM) model to detect HVDs. In [14], the authors extracted the energy and entropy features from each frequency component of the short-time Fourier transform (STFT) synchrosqueezing-based time-frequency-domain (TFD) representation of PCG data and used the SVM classifier to detect HVDs. The machine learning-based methods have shortcomings; for example, they require the evaluation of discriminative features in the time domain, frequency domain, and joint TFD of PCG data. Also, these methods require the selection of features and classifiers for the accurate detection of HVDs. The deep learning methods have advantages such as the automatic feature extraction and classification of HVDs using PCG data. In recent years, various deep learning methods have been reported in the literature to detect HVDs using PCG data. Oh et al. [15] implemented a dilated convolution-based WaveNet deep learning architecture that contains six residual blocks to detect HVDs using PCG data. Baghel et al. [16] proposed a 19-layer-based deep convolutional neural network (CNN) model architecture to detect HVDs using PCG signals. Similarly, in [17], the authors formulated the deep CNN model using the log-mel spectrogram of PCG data to detect HVDs. In [3], the authors implemented a lightweight deep learning model using a polynomial chirplet (PCT)-based TF representation of PCG data to detect HVDs. Though the PCT has better resolution in the TF plane, it has higher computational requirements and requires the selection of the order of the polynomial to approximate the instantaneous frequency for calculating the TF representation of the PCG signal. The other deep learning models have more parameters, and these methods have not been deployed in the cloud-based framework for real-time and IoT-enabled automated detection of HVDs using PCG recordings.

The synchrosqueezed STFT (SSTFT) is an improved and high-resolution-based TFD representation technique for the analysis of nonstationary signals as compared to STFT [18] [14]. TFD analysis of the PCG signal provides more information and can reveal the variations in the heart sound components in both directions (time and frequency). Deep representation learning (DRL) is based on the use of frozen transfer learners to extract representations or features from the image, and the few denser layers after the frozen blocks are used for dimension reduction and classification of images [19]. The advantages of DRL are that these models have fewer training parameters and are based on an unsupervised knowledge transfer mechanism for different image classification applications [19]. The DRL has not been applied in the SSTFT-based joint TFD representation of PCG data to detect HVDs. The novelty of this chapter is the development of a TFD-based DRL model for binary (normal versus abnormal) and multiclass (normal versus AS versus MS versus MR versus MVP) classification of HVDs using

PCG recordings. The remaining sections of this chapter are organized as follows. In Section 10.2, the PCG signal databases used for the evaluation of the TFD-based DRL approach are discussed. The proposed method is described in Section 10.3. The results obtained using this method are presented and discussed in Section 10.4. We have written the conclusion of this chapter in Section 10.5.

10.2 Heart sound signal databases

In this work, we have considered PCG signals from two publicly available databases [20] [21] to assess the HVD diagnosis performance of the proposed TFD-based DRL model. Database 1 contains 1000 PCG recordings, with 200 each belonging to the normal, AS, MR, MS, and MVP classes. Each PCG recording has a sampling frequency of 8 kHz. The duration of the PCG data varies from 2 sec to 2.5 sec with 16 bits as the dynamic range in database 1. Similarly, for database 2, we have used the PCG Signals PhysioNet Challenge 2016 dataset [22] [23], which is available in Kaggle [24], to evaluate the performance of the proposed TFD-based DRL approach. Database 2 consists of 3240 training PCG signal instances and 301 validation PCG instances. The duration of the PCG signal in database 2 varies between 5 sec and 2 min, with a sampling frequency of 2 kHz. Out of 3240 PCG signals, 2575 and 665 PCG signal instances belong to the normal and abnormal classes, respectively [24]. Similarly, the validation set consists of 150 and 151 PCG signal instances for the normal and abnormal classes, respectively [24]. In this work, we have used 665 PCG signal instances from each training set class to avoid the class imbalance problem in the proposed TFD-based DRL approach.

10.3 Proposed method

The flowchart of the proposed TFDDRL approach is depicted in Fig. 10.1. It mainly consists of four essential stages: filtering and amplitude normalization of PCG data, evaluation of the TFD representation using SSTFT, implementation of the DRL-based network, and deployment of the DFDDRL-based method on the cloud-based framework for the automated detection of HVDs. The stages of the flowchart are described in the following subsections.

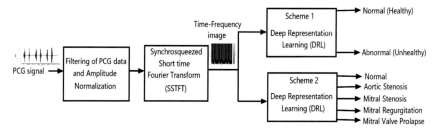

Figure 10.1 Block diagram of the proposed SSTFT-based time-frequency-domain deep representation learning approach to detect HVDs using PCG data.

10.3.1 Filtering and amplitude normalization

There are significant deviations in the spectral characteristics of PCG data in the presence of murmurs and under pathological conditions [3]. In this work, we have applied a bandpass filter with lower and upper cutoff frequency values of 25 Hz and 900 Hz to filter PCG data from both databases [25]. Amplitude normalization is performed by dividing the amplitude value of each sample of PCG data by the maximum amplitude value of the PCG signal for each subject [3]. After filtering and amplitude normalization, all PCG data are subjected to TFD analysis using SSTFT.

10.3.2 Synchrosqueezed short-time Fourier transform

SSTFT is a time-frequency-based reassignment and postprocessing technique to improve the resolution of the STFT-based TFD representation of a nonstationary signal [18]. It is evaluated in three steps. First, for a PCG signal, $z(n)$, the STFT is evaluated as follows [26] [14]:

$$S(n, k) = \sum_{\tilde{n}=0}^{N} z(\tilde{n})w(\tilde{n} - n)e^{\frac{-j2\pi \tilde{n}k}{N}}.$$ (10.1)

After evaluating the time-frequency matrix, $\mathbf{S} = [S(n, k)]_{n,k=0}^{N-1}$, the second step includes the evaluation of the instantaneous frequency, which is given as follows [18]:

$$\mathrm{IF}(n, \tilde{k}) = \mathrm{Real}\{\frac{S(n, \tilde{k}) - S(n - 1, \tilde{k})}{jS(n, \tilde{k})}\},$$ (10.2)

where the numerator is the first derivative of the time-frequency matrix concerning the discrete sample n. In the third step of the SSTFT, the reas-

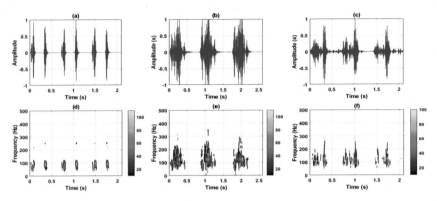

Figure 10.2 (a) Time-domain plot of PCG data for the normal class. (b) Time-domain plot of PCG data for the aortic stenosis (AS) class. (c) Time-domain plot of PCG data for the mitral regurgitation (MR) class. (d) TFD plot of PCG data for the normal class. (e) TFD plot of PCG data for the AS class. (f) TFD plot of PCG data for the MR class.

signment is performed in TFD concerning the new position as $(n, \text{IF}(n, \tilde{k})$. The TFD matrix evaluated using SSTFT is given as follows [27] [14]:

$$\tilde{S}(n, k) = \sum_{\tilde{k}=0}^{N-1} S(n, k)\delta(k - \text{IF}(n, \tilde{k})). \qquad (10.3)$$

The time-domain plots for the PCG signals of the normal, AS, and MR classes are shown in Fig. 10.2(a), Fig. 10.2(b), and Fig. 10.2(c), respectively. Similarly, the SSTFT-based TFD contour plots for PCG signals of the normal, AS, and MR classes are shown in Fig. 10.2(d), Fig. 10.2(e), and Fig. 10.2(f), respectively. It is observed that due to the presence of abnormal heart sound episodes due to AS and MR, significant variations are seen in the TFD representations of PCG signals as compared to the normal case. The TFD representation captures the variations in both time- and frequency-domain characteristics of heart sound due to the presence of pathology. Hence, it captures more information than the time domain of the PCG signal. Therefore, the AI models designed using the joint TFD representation of PCG data are helpful for the accurate detection of HVDs.

10.3.3 TFD-based DRL model

In this work, the DRL model is designed using the TFD representation of the PCG signal. The TFD image of the PCG data is considered, and this

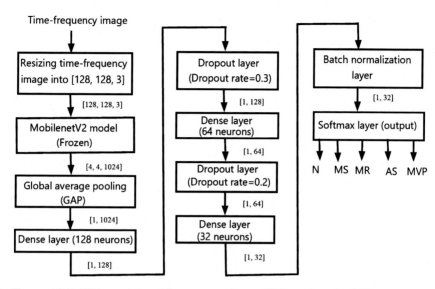

Time-frequency image

Figure 10.3 DRL model architecture to detect HVDs using the TFD representation of PCG data.

image is resized to a tensor with size $128 \times 128 \times 3$. The DRL model architecture to detect HVDs using TFD images of PCG signals is depicted in Fig. 10.3. The main components of the proposed DRL model are the frozen MobileNetV2-based transfer learner block [28], global average pooling (GAP), dense layers, drop-out layers, batch normalization, and an output layer, respectively. For the TFD image of the PCG signal with size $128 \times 128 \times 3$, the tensor obtained after the frozen MobileNetV2-based transfer learner block has the shape of $\mathcal{A} \in 4 \times 4 \times 1024$ [28]. We have applied the GAP layer after the frozen MobileNetV2 layer, and mathematically, the output of the GAP layer is defined as follows [29] [19]:

$$b(1, q) = \frac{1}{MN} \sum_{i=1}^{M} \sum_{j=1}^{N} A(i, j, q), \tag{10.4}$$

where $\mathbf{b} = [b(1, q)]_{q=1}^{Q}$. Here, the values of M, N, and Q are 4, 4, and 1024, respectively. Similarly, for the dense layer, the feature vector at the lth layer is evaluated as follows [30]:

$$\mathbf{z}^l = f(\mathbf{W}^l \mathbf{z}^l + bl), \tag{10.5}$$

where $\mathbf{z}^l$ and $\mathbf{z}^{l-1}$ are the feature vectors for the lth and $(l-1)$th hidden layer, respectively. Similarly, $\mathbf{W}^l$ and bl are the weight matrix and bias vector corresponding to the transition from the $(l-1)$th to the lth hidden layer [30]. Likewise, we have used the batch normalization layer before the output layer, and the mathematical expression for this layer is given as follows [31]:

$$\tilde{\mathbf{c}} = \mathrm{sc} \times \hat{\mathbf{c}} + \mathrm{sh}, \tag{10.6}$$

where sc and sh are interpreted as the scale and shift factors in the batch normalization layer, respectively. Similarly, the factor $\hat{\mathbf{c}}$ is defined as follows [31]:

$$\hat{\mathbf{c}} = \frac{(\mathbf{c} - \mu_K)}{\sqrt{\sigma_K^2 + \epsilon}}, \tag{10.7}$$

where μ_K and σ_K^2 are the mean and variance values for a minibatch size of K. For the PCG signals from database 1, we have five classes (normal and four HVD classes), and hence the DRL model contains five output neurons. Similarly, for the PCG signals from database 2, the two class–based HVD detection tasks, such as normal versus abnormal, are formulated. The DRL model designed for database 2 has two output neurons representing the normal and abnormal classes, respectively. The training and test PCG signal instances for the proposed TFD-based DRL model are selected using hold-out validation and 5-fold cross-validation (CV)-based methodologies [32]. The hyperparameters of the DRL model include a batch size of 32, a number of epochs of 10, and an initial learning rate of 0.01. The cost function and the optimizer are used as categorical cross-entropy and adaptive moment (ADAM) [33] for the multiclass classification of HVDs using the DRL model. Similarly, the binary cross-entropy-based cost function is used for the two-class-based classification of PCG data using the proposed TFD-based DRL model. Performance metrics such as precision, accuracy, F1-score, recall, and the kappa score [34] [3] are used to measure the classification performance of the DRL model to detect HVDs.

10.4 Results and discussion

We have shown the results by evaluating the proposed time-frequency-based DRL model using PCG data from both databases. Table 10.1 shows the SSTFT-domain DRL approach's HVD detection results for hold-out validation. It is noted that for database 1 PCG signals, the SSTFT-domain

Table 10.1 Performance of the SSTFT-domain DRL model for heart valve disease detection using hold-out validation.

Database	Classes	Accuracy (%)	Precision (%)	Recall (%)	F1-score (%)	Kappa (%)
Database 1	5	98.00	98.00	98.30	98.10	97.50
Database 2	2	90.60	88.62	84.38	86.22	72.50

Table 10.2 Performance of the SSTFT-domain DRL model for heart valve disease detection using 5-fold cross-validation.

Database	Classes	Accuracy (%)	Precision (%)	Recall (%)	F1-score (%)	Kappa (%)
Database 1	5	96.60 ± 0.97	96.71 ± 0.90	96.60 ± 0.97	96.58 ± 0.98	95.74 ± 1.21
Database 2	2	87.56 ± 3.14	86.29 ± 2.09	77.78 ± 8.96	79.34 ± 8.79	59.67 ± 15.82

DRL model has demonstrated an accuracy value of 98% to detect and classify five classes (four HVD classes and one normal class) using PCG data. Similarly, the precision, recall, and F1-score values are also obtained as more than 98% using the SSTFT-domain DRL model. The classification accuracy of the SSTFT-domain DRL is obtained as 90.60% using PCG signals from database 2. Table 10.2 shows the results evaluated using the SSTFT-domain DRL approach using PCG data from both databases with the 5-fold CV. It is observed that for database 1, the SSTFT-domain DRL has produced average values of performance measures (accuracy, precision, recall, and F1-score) along all 5-fold CVs of more than 96% for five-class-based HVD detection using PCG data. Similarly, for the two-class-based classification task using PCG signals from database 2, the DRL model obtained an average accuracy value of more than 87% using 5-fold CV. The kappa score value is less for the database 2 case than the database 1 case evaluated using the SSTFT-domain DRL model using PCG signals. The class imbalance case of the proposed SSTFT-domain DRL model is observed for the database 2 case during the testing phase. Hence, a lower kappa value is obtained for database 2 using the SSTFT-domain DRL model to detect HVDs. The performance of DRL is evaluated using different window sizes of the SSTFT and STFT-based time-frequency representation of PCG signals with Hanning windows from database 1 to detect HVDs; these results are shown in Table 10.3. For a window size of 512, the accuracy values of DRL are obtained as 97.50% and 98.00%, using an STFT- and an SSTFT-

Table 10.3 Accuracy of DRL versus window size for SSTFT for heart valve disease detection using hold-out validation.

Time-frequency method	Window size	Accuracy (%)
STFT	16	91.80
STFT	32	94.50
STFT	64	93.00
STFT	128	95.50
STFT	256	96.00
STFT	512	97.50
SSTFT	16	95.50
SSTFT	32	95.50
SSTFT	64	96.50
SSTFT	128	95.50
SSTFT	256	98.00
SSTFT	512	98.00

Table 10.4 Comparison with other time-frequency-domain DRL methods for HVD detection with hold-out validation.

Time-frequency method	Accuracy (%)
MFCC	91.50
Stockwell Transform	79.00
STFT	97.50
SSTFT	98.00

based time-frequency representation of PCG signals. The DRL coupled with SSTFT with a window size of 512 has demonstrated higher accuracy compared to other window sizes. Similarly, we have also evaluated the performance of DRL using other window-based SSTFT-domain representations of PCG signals to detect HVDs. The Blackmann and Hamming window–based SSTFT and DRL models have lower classification accuracy in detecting HVDs than the Hanning window-based SSTFT and DRL models.

Furthermore, we have evaluated the performance of the SSTFT and other time-frequency analysis–based methods with DRL to detect HVDs; the results are depicted in Table 10.4. It is evident that the SSTFT has produced higher classification accuracy in detecting HVDs than MFCC,

Table 10.5 Classification performance of SSTFT-domain DRL using posttraining quantization of model weight matrices, bias vectors, and activation functions to detect HVDs using PCG signals from database 1.

Quantization level	Accuracy (%)	Precision (%)	Recall (%)	F1-score (%)	Kappa (%)
FP32	98.00	98.00	98.30	98.10	97.50
FP16	96.88	96.60	96.65	96.58	96.06
INT8	26.25	6.89	26.25	10.92	4.62

Table 10.6 Classification performance of SSTFT-domain DRL using posttraining quantization of model weight matrices, bias vectors, and activation functions for classifying normal and abnormal classes using PCG signals from database 2.

Quantization level	Accuracy (%)	Precision (%)	Recall (%)	F1-score (%)	Kappa (%)
FP32	90.60	88.62	84.38	86.22	72.50
FP16	79.73	83.56	79.79	79.16	59.51
INT8	49.84	24.83	49.84	33.15	5.08

Stockwell transform, and STFT-based methods with DRL. We have considered the quantization of the DRL model parameters using FP16 and INT8-based quantization cases. The quantization of the DRL models to detect HVDs using TFD analysis of PCG signals from database 1 and database 2 are depicted in Table 10.5 and Table 10.6, respectively. The INT8 and FP16 quantization cases help reduce the DRL model's size. However, the accuracy of the DRL model is reduced for both database 1 and database 2 for five-class- and two-class-based automated detection HVDs using SSTFT-based TFD representation of PCG data.

Moreover, we have deployed the SSTFT-domain DRL model on the streamlit cloud-based framework for real-time detection of HVDs using heart sound recordings from both databases. For database 1, the inferences of the proposed SSTFT-domain DRL models on the cloud-based framework to detect normal, AS, MS, MR, and MVP classes using PCG recordings are shown in Fig. 10.4(a), Fig. 10.4(b), Fig. 10.4(c), Fig. 10.4(d), and Fig. 10.4(e), respectively. It is observed that for all these five PCG recordings, the method (SSTFT-domain DRL) deployed on the cloud has predicted the normal class and all four HVD classes correctly. Similarly, for database 2, the inferences of the proposed SSTFT-domain DRL models on

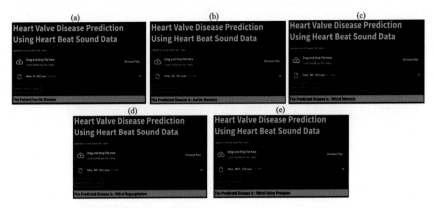

Figure 10.4 (a) Inference of SSTFT-domain DRL model on streamlit-based cloud framework to predict the healthy subjects (normal class) using PCG data. (b) Inference of SSTFT-domain DRL model on streamlit-based cloud framework to detect AS using PCG data. (c) Inference of SSTFT-domain DRL model on streamlit-based cloud framework to detect MS using PCG data. (d) Inference of SSTFT-domain DRL model on streamlit-based cloud framework to detect MR using PCG data. (e) Inference of SSTFT-domain DRL model on streamlit-based cloud framework to detect MVP using PCG data.

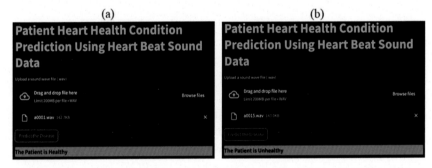

Figure 10.5 (a) Deployment of SSTFT-domain DRL on the cloud framework to identify healthy subjects (normal class) using PCG data. (b) Deployment of SSTFT-domain DRL on the cloud framework to identify unhealthy subject (abnormal classes) using PCG data.

the cloud–based framework to detect normal and abnormal classes using PCG recordings are shown in Fig. 10.5(a) and Fig. 10.5(b), respectively. Normal and abnormal classes are successfully predicted using the proposed SSTFT–domain DRL model when inference is performed on the cloud–based framework.

Table 10.7 Comparison with different existing methods to detect HVDs using PCG signals from database 1 and database 2.

Methods considered	Database used	Overall accuracy (%)
Features extracted using MFCC and DWT-based methods with PCG data and SVM classifier [20]	Database 1	97.90
Magnitude and phase features extracted using wavelet synchrosqueezing transform of PCG data and random forest classifier [9]	Database 1	95.13
Fourier synchrosqueezing-domain energy and entropy features evaluated from PCG and SVM classifier [14]	Database 1	96.33
Dilated 1D convolution-based residual CNN model with PCG data as input [15]	Database 1	97.00
Triple spectrogram-based TFD representation of PCG data and attention-based deep learning model [35]	Database 1	97.77
The time interval, MFCC, and DTW-based features extracted from PCG data and SVM classifier [13]	Database 2	82.40
Heatmap-based images obtained from the MFCC of PCG data and deep CNN model [36]	Database 2	84.80
STFT-based TFD images evaluated from PCG signals and deep CNN model [37]	Database 2	81.30
CNN and bidirectional LSTM-based hybrid deep learning model with PCG data as input [38]	Database 2	87.31
TFD images evaluated using frequency-domain polynomial chirplet transform of PCG data and deep CNN model [3]	Database 2	85.16
SSTFT-domain time-frequency images evaluated from PCG data and DRL model (present work)	Database 1 and database 2	98.00 and 90.60

Furthermore, we have compared the HVD detection performance of the proposed SSTFT-domain DRL model with various existing techniques using PCG data from both database 1 and database 2. The comparison results are shown in Table 10.7. It is observed that the machine learning-based methods [20] [9] [14] [13] coupled with different transform-domain features of PCG data have demonstrated lower overall accuracy for both multiclass and binary HVD schemes using database 1 and database 2. Similarly, the deep learning models applied in the MFCC, STFT, and polynomial chirplet transform-domain time-frequency analysis of PCG data have obtained lower overall classification accuracy than the proposed approach to detect HVDs. In [15] and [38], the authors utilized PCG data directly as input to deep learning models and obtained overall accuracy values of 97% and 87.31%, respectively. However, the proposed SSTFT-domain DRL model has produced higher accuracy than these methods using PCG data from both databases. Implementing and developing an IoT-enabled healthcare framework to detect HVDs using PCG recordings is challenging. The proposed TFD-based DRL model has been deployed on the cloud-based framework, and the inference of the model is performed using test PCG recordings from different subjects. The heart sound or PCG data can be recorded from the subjects using a digital stethoscope [39]. Once the PCG data have been recorded, they can be transmitted to the cloud using either Bluetooth and Wifi or a long-range, wide-area network (LoRaWAN) gateway [1]. In this work, we have prepared a web application where the inference of the model can be performed using Android devices. This study tests the proposed TFD-based DRL model using PCG signals from two different databases. In an IoT-based healthcare system, the AI model can successfully work on PCG recordings from the subjects of one hospital with higher accuracy. However, when the same AI model's inference is performed on the PCG recordings from the subjects of another hospital, the model may fail to produce higher classification performance to detect HVDs. Hence, it is interesting to develop a federated learning-based IoT-enabled healthcare system [40] that provides an optimized AI model to obtain higher accuracy in detecting HVDs using PCG signals recorded from the subjects in different hospitals. Privacy and security of patient data and AI model parameters are crucial for implementing IoT-enabled healthcare systems to detect HVDs. The inference time or latency is high when the inference of the TFD-based DRL model is performed using the cloud-based framework to detect HVDs using PCG recordings. Hence, there is a requirement to process PCG data locally (nearest to the recording device or digital stethoscope) or use edge computing-based devices to detect HVDs. There is a potential to explore low-power edge devices such as microcontrollers and field programmable gate arrays (FPGAs) [41] for real-time detection of HVDs using PCG sensor data.

10.5 Conclusion

The TFD-based DRL approach has been proposed in this chapter for the automated detection of HVDs using PCG recordings. The performance of this approach has been evaluated using binary classes (normal versus abnormal) and five classes (normal versus AS versus MS versus MR versus MVP) based on HVD detection using PCG data. The suggested approach has demonstrated classification accuracy values of more than 90% for both classification schemes using PCG signals from two databases with hold-out validation. The proposed approach has been deployed on the cloud-based framework for real-time HVD detection using PCG recordings. The TFD-based DRL approach is simple and cost-effective to detect HVDs using PCG signals. Hence, it can be used for IoT-enabled healthcare monitoring using heart sound data recorded from the subjects using a digital stethoscope.

References

[1] Z.N. Aghdam, A.M. Rahmani, M. Hosseinzadeh, The role of the Internet of things in healthcare: future trends and challenges, Computer Methods and Programs in Biomedicine 199 (2021) 105903.

[2] S. Cao, X. Lin, K. Hu, L. Wang, W. Li, M. Wang, Y. Le, Cloud computing-based medical health monitoring IoT system design, Mobile Information Systems 2021 (2021) 1–12.

[3] J. Karhade, S. Dash, S.K. Ghosh, D.K. Dash, R.K. Tripathy, Time–frequency-domain deep learning framework for the automated detection of heart valve disorders using PCG signals, IEEE Transactions on Instrumentation and Measurement 71 (2022) 1–11.

[4] S.K. Ghosh, R. Ponnalagu, R. Tripathy, U.R. Acharya, Automated detection of heart valve diseases using chirplet transform and multiclass composite classifier with PCG signals, Computers in Biology and Medicine 118 (2020) 103632.

[5] Y. Sato, T. Kawasaki, S. Honda, K. Harimoto, S. Miki, T. Kamitani, H. Shiraishi, S. Matoba, Third and fourth heart sounds and myocardial fibrosis in hypertrophic cardiomyopathy, Circulation Journal 82 (2) (2018) 509–516.

[6] C. Spiers, Cardiac auscultation, British Journal of Cardiac Nursing 6 (10) (2011) 482–486.

[7] M. El-Segaier, O. Lilja, S. Lukkarinen, L. Sörnmo, R. Sepponen, E. Pesonen, Computer-based detection and analysis of heart sound and murmur, Annals of Biomedical Engineering 33 (2005) 937–942.

[8] S.I. Khan, S.M. Qaisar, R.B. Pachori, Automated classification of valvular heart diseases using FBSE-EWT and PSR based geometrical features, Biomedical Signal Processing and Control 73 (2022) 103445.

[9] S.K. Ghosh, R.K. Tripathy, R. Ponnalagu, R.B. Pachori, Automated detection of heart valve disorders from the PCG signal using time-frequency magnitude and phase features, IEEE Sensors Letters 3 (12) (2019) 1–4.

[10] S.K. Ghosh, R. Ponnalagu, R. Tripathy, U.R. Acharya, et al., Deep layer kernel sparse representation network for the detection of heart valve ailments from the time-frequency representation of PCG recordings, BioMed Research International 2020 (2020).

[11] T. Tuncer, S. Dogan, R.-S. Tan, U.R. Acharya, Application of Petersen graph pattern technique for automated detection of heart valve diseases with PCG signals, Information Sciences 565 (2021) 91–104.

[12] P. Langley, A. Murray, Abnormal heart sounds detected from short duration unsegmented phonocardiograms by wavelet entropy, in: 2016 Computing in Cardiology Conference (CinC), IEEE, 2016, pp. 545–548.

[13] J.J.G. Ortiz, C.P. Phoo, J. Wiens, Heart sound classification based on temporal alignment techniques, in: 2016 Computing in Cardiology Conference (CinC), IEEE, 2016, pp. 589–592.

[14] S.K. Ghosh, R.K. Tripathy, R. Ponnalagu, Classification of PCG signals using Fourier-based synchrosqueezing transform and support vector machine, in: 2021 IEEE Sensors, IEEE, 2021, pp. 1–4.

[15] S.L. Oh, V. Jahmunah, C.P. Ooi, R.-S. Tan, E.J. Ciaccio, T. Yamakawa, M. Tanabe, M. Kobayashi, U.R. Acharya, Classification of heart sound signals using a novel deep WaveNet model, Computer Methods and Programs in Biomedicine 196 (2020) 105604.

[16] N. Baghel, M.K. Dutta, R. Burget, Automatic diagnosis of multiple cardiac diseases from PCG signals using convolutional neural network, Computer Methods and Programs in Biomedicine 197 (2020) 105750.

[17] M.T. Nguyen, W.W. Lin, J.H. Huang, Heart sound classification using deep learning techniques based on log-mel spectrogram, Circuits, Systems, and Signal Processing 42 (1) (2023) 344–360.

[18] T. Oberlin, S. Meignen, V. Perrier, The Fourier-based synchrosqueezing transform, in: 2014 IEEE International Conference on Acoustics, Speech and Signal Processing (ICASSP), IEEE, 2014, pp. 315–319.

[19] S. Bhaskarpandit, A. Gade, S. Dash, D.K. Dash, R.K. Tripathy, R.B. Pachori, Detection of myocardial infarction from 12-lead ECG trace images using eigendomain deep representation learning, IEEE Transactions on Instrumentation and Measurement 72 (2023) 1–12.

[20] Yaseen, G.-Y. Son, S. Kwon, Classification of heart sound signal using multiple features, Applied Sciences 8 (12) (2018) 2344.

[21] yaseen21khan, Yaseen21khan/classification-of-heart-sound-signal-using-multiple-features-: data plus code of classification of heart sound signal using multiple features, https://github.com/yaseen21khan/Classification-of-Heart-Sound-Signal-Using-Multiple-Features-.

[22] G.D. Clifford, C. Liu, B. Moody, D. Springer, I. Silva, Q. Li, R.G. Mark, Classification of normal/abnormal heart sound recordings: the PhysioNet/Computing in Cardiology Challenge 2016, in: 2016 Computing in Cardiology Conference (CinC), IEEE, 2016, pp. 609–612.

[23] A.L. Goldberger, L.A. Amaral, L. Glass, J.M. Hausdorff, P.C. Ivanov, R.G. Mark, J.E. Mietus, G.B. Moody, C.-K. Peng, H.E. Stanley, PhysioBank, PhysioToolkit, and PhysioNet: components of a new research resource for complex physiologic signals, Circulation 101 (23) (2000) e215–e220.

[24] Swapnil, Heart sound database, https://www.kaggle.com/datasets/swapnilpanda/heart-sound-database, Jul 2020.

[25] S.K. Ghosh, R. Ponnalagu, R.K. Tripathy, G. Panda, R.B. Pachori, Automated heart sound activity detection from PCG signal using time–frequency-domain deep neural network, IEEE Transactions on Instrumentation and Measurement 71 (2022) 1–10.

[26] L. Durak, O. Arikan, Short-time Fourier transform: two fundamental properties and an optimal implementation, IEEE Transactions on Signal Processing 51 (5) (2003) 1231–1242.

[27] S. Madhavan, R.K. Tripathy, R.B. Pachori, Time-frequency domain deep convolutional neural network for the classification of focal and non-focal EEG signals, IEEE Sensors Journal 20 (6) (2019) 3078–3086.

[28] M. Sandler, A. Howard, M. Zhu, A. Zhmoginov, L.-C. Chen, MobileNetV2: inverted residuals and linear bottlenecks, in: Proceedings of the IEEE Conference on Computer Vision and Pattern Recognition, 2018, pp. 4510–4520.

[29] Z. Li, S.-H. Wang, R.-R. Fan, G. Cao, Y.-D. Zhang, T. Guo, Teeth category classification via seven-layer deep convolutional neural network with max pooling and global average pooling, International Journal of Imaging Systems and Technology 29 (4) (2019) 577–583.

[30] R. Tripathy, U.R. Acharya, Use of features from RR-time series and EEG signals for automated classification of sleep stages in deep neural network framework, Biocybernetics and Biomedical Engineering 38 (4) (2018) 890–902.

[31] N. Bjorck, C.P. Gomes, B. Selman, K.Q. Weinberger, Understanding batch normalization, Advances in Neural Information Processing Systems 31 (2018).

[32] T. Siddharth, R.K. Tripathy, R.B. Pachori, Discrimination of focal and non-focal seizures from EEG signals using sliding mode singular spectrum analysis, IEEE Sensors Journal 19 (24) (2019) 12286–12296.

[33] I. Goodfellow, Y. Bengio, A. Courville, Deep Learning, MIT Press, 2016.

[34] R. Tripathy, S. Dandapat, Detection of cardiac abnormalities from multilead ECG using multiscale phase alternation features, Journal of Medical Systems 40 (2016) 1–9.

[35] S. Chowdhury, M. Morshed, S.A. Fattah, SpectroCardioNet: an attention-based deep learning network using triple-spectrograms of PCG signal for heart valve disease detection, IEEE Sensors Journal 22 (23) (2022) 22799–22807.

[36] J. Rubin, R. Abreu, A. Ganguli, S. Nelaturi, I. Matei, K. Sricharan, Classifying heart sound recordings using deep convolutional neural networks and mel-frequency cepstral coefficients, in: 2016 Computing in Cardiology Conference (CinC), IEEE, 2016, pp. 813–816.

[37] T. Nilanon, J. Yao, J. Hao, S. Purushotham, Y. Liu, Normal/abnormal heart sound recordings classification using convolutional neural network, in: 2016 Computing in Cardiology Conference (CinC), IEEE, 2016, pp. 585–588.

[38] M. Alkhodari, L. Fraiwan, Convolutional and recurrent neural networks for the detection of valvular heart diseases in phonocardiogram recordings, Computer Methods and Programs in Biomedicine 200 (2021) 105940.

[39] A. Jain, R. Sahu, A. Jain, T. Gaumnitz, P. Sethi, R. Lodha, Development and validation of a low-cost electronic stethoscope: DIY digital stethoscope, BMJ Innovations 7 (4) (2021) bmjinnov-2021.

[40] S. Singh, S. Rathore, O. Alfarraj, A. Tolba, B. Yoon, A framework for privacy-preservation of IoT healthcare data using federated learning and blockchain technology, Future Generation Computer Systems 129 (2022) 380–388.

[41] T. Belabed, M.G.F. Coutinho, M.A. Fernandes, C.V. Sakuyama, C. Souani, User driven FGPA-based design automated framework of deep neural networks for low-power low-cost edge computing, IEEE Access 9 (2021) 89162–89180.

Index

L

Least mean square (LMS) approach, 100
Left atrium (LA), 1
Left ventricle (LV), 1, 82
Light-emitting diode (LED) transmitter, 11
Limitations, 58
Linear discriminant analysis (LDA), 70
Linear prediction coefficient (LPC), 10
Locality-preserving projection (LPP), 128
Logistic regression models, 122
Long-range, wide-area network
(LoRaWAN) gateway, 162
Low-arousal (LA), 102, 110
Low-frequency component (LFC), 87
Low-frequency noise filtering, 112
Low-pass filter, 23

M

Machine learning (ML), 7, 66, 72, 143
 algorithms, 36, 72, 140, 142, 143
 approaches, 66, 150
 classifiers, 83
 incorporation, 55
 methods, 5
 model, 55, 71–73, 122
 techniques, 20, 70, 76, 150
Male frequency cepstral coefficient
 (MFCC), 150, 158, 162
Matthews correlation coefficient (MCC),
 111
 average, 115
 values, 111, 113, 114
Maximal overlap discrete wavelet transform
 (MODWT), 12
Mean arterial pressure (MAP), 135
Mean blood pressure (MBP), 135
Mean square error (MSE), 100
Median filter, 108, 109, 111, 112, 115
Mel frequency cepstral coefficient
 (MFCC), 10
 features, 84, 85
Mental health, 65, 67, 69
 condition, 67, 68, 77
 issues, 67, 69
 problems, 67, 69, 77
 status, 77

Mitral regurgitation (MR), 8, 150, 154
Mitral stenosis (MS), 8, 150
Mitral valve (MV), 1
Mobile Atrial Fibrillation Application
 (MAFA), 54
Mobius function, 39
Model
 classification, 73, 83, 85, 94
 deep learning, 10, 36, 37, 55, 84, 85, 94,
 151, 162
 machine learning, 55, 71–73, 122
Modified frequency slice wavelet transform
 (MFSWT), 93
Morphological
 changes, 3
 features, 7, 114, 115, 122, 131
 features ECG data, 6, 7
Mother wavelet function, 126
Motion artifacts, 58, 100, 102, 108, 144
Multichannel
 deep CNN, 37, 40, 43, 45
 accuracy, 42
 architecture, 42, 43, 45
 classification results, 42
 ECG signals, 32
 EEG signals, 20
 signals, 24
Multidomain datasets, 91
Multilayer perceptron (MLP), 122
Multilead
 beat, 22, 23, 26, 32
 ECG signals, 21, 22
Multilead fusion (MLF) method, 23, 30–32
Multisite PPG
 measurement, 144
 signals, 145
Multivariate fast and adaptive empirical
 mode decomposition (MFAEMD),
 13
Multivariate projection-based fixed
 boundary empirical wavelet
 transform (MPFBEWT), 20, 21,
 24, 26, 32
 algorithm, 21
 filter bank, 24, 25
 technique, 28
Multivariate signals, 20, 31

Printed in the United States
by Baker & Taylor Publisher Services